Science and Technology for Water Purification

Science and Technology for Water Purification

Guest Editors

Yuanfeng Qi
Kai He
Jing Wei
Zhishui Liang

MDPI

Basel • Beijing • Wuhan • Barcelona • Belgrade • Novi Sad • Cluj • Manchester

Guest Editors

Yuanfeng Qi
School of Environmental and Municipal Engineering
Qingdao University of Technology
Qingdao
China

Kai He
School of Civil Engineering
Sun Yat-sen University
Zhuhai
China

Jing Wei
School of the Environment and Safety Engineering
Jiangsu University
Zhenjiang
China

Zhishui Liang
School of Civil Engineering
Southeast Univeristy
Nanjing
China

Editorial Office
MDPI AG
Grosspeteranlage 5
4052 Basel, Switzerland

This is a reprint of the Special Issue, published open access by the journal *Water* (ISSN 2073-4441), freely accessible at: https://www.mdpi.com/journal/water/special_issues/76MQ2F96O6.

For citation purposes, cite each article independently as indicated on the article page online and as indicated below:

Lastname, A.A.; Lastname, B.B. Article Title. *Journal Name* **Year**, *Volume Number*, Page Range.

ISBN 978-3-7258-5051-8 (Hbk)
ISBN 978-3-7258-5052-5 (PDF)
https://doi.org/10.3390/books978-3-7258-5052-5

Contents

About the Editors

Yuanfeng Qi

Yuanfeng Qi (Qingdao University of Technology) focuses on wastewater treatment, specializing in waste-to-resource recycling, inorganic membrane investment and reactor design. He develops novel fillers and carriers for pollutant removal and pioneers low-carbon infrastructure models. His methodologies for evaluating wastewater-induced environmental damage integrate life-cycle analysis with risk modeling, guiding industrial remediation.

Kai He

Kai He (Sun Yat-sen University) conducts research in water-quality monitoring and environmental analytics. Drawing on experience in several international research groups, he explores practical approaches to contemporary water-engineering challenges. His work centers on three areas: (i) combining spectroscopic and mass-spectrometric techniques with large-scale data analysis, (ii) tracing pollution sources in aquatic environments, and (iii) developing instruments for real-time environmental monitoring. He has contributed to methods for assessing pharmaceutical residues through wastewater surveillance and to machine-learning models that estimate the formation of disinfection by-products.

Jing Wei

Jing Wei (Jiangsu University) specializes in energy-efficient membrane technologies and water treatment. She designs high-flux membranes for simultaneous pollutant degradation and clean energy generation. Her modular biofilm reactors enable decentralized wastewater treatment, while her microbial carrier research aims to achieve enganced nitrogen removal and low energy consumption in wastewater treatment plants.

Zhishui Liang

Zhishui Liang (Southeast University) integrates ecological remediation with water treatment, developing nature-based solutions like constructed wetland–bioreactor hybrids. His phytoremediation research uses hyperaccumulator plants for heavy metal removal, and his current projects explore microbial–electrochemical interfaces to optimize agricultural runoff treatment and groundwater recharge.

water

MDPI

Editorial

Science and Technology for Water Purification: Achievements and Strategies

Yuanfeng Qi [1] and Kai He [2,*]

1 School of Environmental and Municipal Engineering, Qingdao University of Technology, Qingdao 266033, China; qiyuanfeng@qut.edu.cn
2 School of Civil Engineering, Sun Yat-sen University, No. 2 University Road, Zhuhai 519082, China
* Correspondence: hekai7@mail.sysu.edu.cn

1. Introduction

Water scarcity and pollution remain two of the most pressing global challenges, with profound implications for public health, environmental sustainability, and economic development [1,2]. As the global population continues to grow and industrial activities expand, the demand for clean, safe water is steadily increasing [3]. However, water bodies are increasingly polluted by a variety of contaminants, including industrial chemicals [4], agricultural runoff [3,5], pharmaceuticals [4], and microplastics [6]. These pollutants complicate traditional water treatment processes and present significant challenges [7–9], as many emerging contaminants are difficult to remove using conventional filtration and chemical treatments. Consequently, there is a pressing need for more advanced and efficient water purification technologies capable of addressing these evolving issues.

In addition to the growing problem of pollution, many regions are already facing water scarcity, making it crucial to not only focus on wastewater purification but also prioritize its reuse and resource recovery [10]. Traditional purification methods, such as coagulation, filtration, and chemical treatments, often prove inadequate when dealing with the complex nature of modern wastewater [11]. The presence of emerging pollutants such as endocrine disruptors, pharmaceuticals, and novel chemicals further complicates treatment efforts [12]. Therefore, there is an urgent demand for innovative, cost-effective, and sustainable water treatment technologies that can effectively address these new challenges while improving the overall efficiency, sustainability, and affordability of water purification systems [13].

This Special Issue, "Science and Technology for Water Purification", brings together cutting-edge research on the latest advancements in water and wastewater treatment technologies, with a focus on innovative solutions that address the multifaceted challenges of water pollution and scarcity. As global water demand increases and water quality deteriorates due to the discharge of complex pollutants, the research highlighted in this Special Issue offers new perspectives on how to improve existing water treatment processes while creating more sustainable and efficient alternatives. One of the primary goals is to explore innovations aimed at enhancing pollutant removal efficiency, reducing energy consumption, and improving resource recovery from wastewater. The contributions cover a broad spectrum of technological advancements, including the development of novel materials such as advanced adsorbents, catalysts, and nanomaterials that offer superior performance in removing a wide range of contaminants, including heavy metals, pharmaceutical residues, and microplastics. These materials are central to addressing the limitations of conventional treatment methods, which often struggle to effectively remove emerging pollutants. Additionally, this Special Issue features research on innovative reactor designs and hybrid treatment systems that can increase the effectiveness and efficiency of wastewater treatment, particularly in high-pollution environments. Another significant focus is the integration of process management strategies and eco-friendly engineering approaches that contribute to sustainable water treatment. These strategies emphasize the need for systems that not only remove contaminants but also recover valuable resources, such as nutrients (e.g., nitrogen and phosphorus) and energy, from wastewater. By incorporating these resource recovery principles, the papers demonstrate how wastewater treatment can evolve from a purely waste management function to a more sustainable, circular process that supports both environmental and economic goals.

2. Review of New Advances

2.1. Reviews on Science and Technology for Water Purification

Two comprehensive review articles are contributed to this Special Issues and survey the latest research and advancements in the field of water and wastewater treatment, providing valuable insights into emerging technologies, trends, and future directions in water purification science. By offering an in-depth analysis of the current state-of-the-art methods, these reviews set the stage for the next wave of innovations in water treatment technologies.

Wang et al. (contribution 1) provide a comprehensive synthesis of research on the removal of contaminants from wastewater using constructed wetland (CW) substrates. The review highlights the role of CW substrates as the primary medium supporting microbial communities, plant roots, and other organisms involved in contaminant removal. It evaluates various substrate types—organic, inorganic, and composite materials—and assesses their effectiveness in treating a wide range of contaminants. Specifically, the following key findings are extracted:

(a) The review categorizes CW substrates, outlining the advantages and limitations of organic, inorganic, and composite materials. Organic substrates, such as compost and peat, provide nutrients for microbes but have limited adsorption capacity. Inorganic materials like sand and gravel offer better physical support and filtration but less microbial activity. Composite substrates, combining organic and inorganic components, enhance both contaminant removal and microbial support.

(b) The primary mechanisms of contaminant removal in CW systems are elucidated, including physical filtration, adsorption, and bioremediation. The combination of physical and biological processes allows CW systems to remove a wide range of pollutants, such as heavy metals, organic compounds, nutrients (nitrogen and phosphorus), and pathogens. Substrate properties, such as porosity and surface area, are shown to influence filtration and adsorption efficiency.

(c) The performance of CW systems in contaminant removal is evaluated, and the effectiveness of different substrates in laboratory and field-scale experiments is compared. The review indicates that selecting and optimizing suitable substrates can lead to high removal efficiencies for various contaminants.

(d) The review addresses challenges, such as substrate clogging, contaminant leaching, and the need for long-term substrate stability under variable conditions. The authors suggest that future research should focus on developing more durable, efficient, and environmentally friendly substrates and optimizing CW system designs to improve performance and reduce maintenance.

(e) The environmental and economic advantages of CW systems for wastewater treatment are emphasized. There is no doubt that wetlands are cost-effective, low-maintenance, and energy-efficient compared to conventional technologies. Multiple ecosystem services are pointed out, such as wildlife habitat, carbon sequestration, and nutrient cycling. The authors advocate broader adoption of CW systems, particularly in small communities, rural areas, and developing countries where conventional treatment infrastructure may be lacking.

Overall, this review provides significant contributions to the field of wastewater treatment by offering a comprehensive analysis of the role of CW substrates in the removal of contaminants. The paper highlights the potential of constructed wetlands as an eco-friendly and cost-effective solution for wastewater treatment, particularly in settings where conventional treatment infrastructure is lacking or unaffordable. By identifying the strengths and limitations of various substrate materials, the review offers valuable guidance for future research and the design of more efficient and sustainable CW systems.

It is known that microplastics have emerged as a significant environmental pollutant and are widely distributed in aquatic ecosystems [6]. Yan et al. (contribution 2) described the interaction between microplastics (MPs) and dissolved organic compounds (DOCs) in water environments. Dissolved organic compounds, including natural and synthetic organic chemicals, are prevalent in water due to industrial, agricultural, and domestic activities. The study focuses on the pathways through which increasing amounts of MPs in water bodies influence the behavior and fate of DOCs and provides insights into the coupled effects of these two types of contaminants on water quality and treatment processes. To be specific, the core findings are as follows:

(a) The study demonstrates that microplastics can adsorb various dissolved organic compounds, particularly hydrophobic DOCs, due to the large surface area and chemical properties of MPs. This interaction leads to changes in the distribution and concentration of DOCs in aquatic environments, potentially altering the behavior and bioavailability of organic pollutants. The authors highlight that MPs serve as a carrier for organic contaminants, thus enhancing their transport and persistence in water systems.

(b) The paper discusses how different characteristics of MPs, such as size, shape, and surface chemistry, influence the adsorption of DOCs. Smaller MPs with larger surface areas show a higher capacity for adsorbing DOCs, while those with functionalized surfaces tend to adsorb specific types of organic compounds more effectively. This finding underscores the complexity of interactions between MPs and DOCs, which can vary depending on the physical and chemical properties of the MPs present.

(c) The study addresses the coupling relationships between MPs and DOCs for aquatic ecosystems and water quality management. The presence of MPs varies the natural cycling of organic matter, potentially leading to the accumulation of harmful organic contaminants in sediments or biota. The adsorption of DOCs on MPs might influence microbial communities in aquatic ecosystems, as the bioavailability of organic compounds can be modified by their association with microplastics.

(d) The interactions between MPs and DOCs that could affect water treatment processes are complex. Traditional water treatment methods may be less effective in removing MPs and the organic contaminants they carry. The review highlights the need for developing more advanced treatment technologies capable of addressing both microplastic pollution and the complex mixtures of organic pollutants that may be adsorbed onto MPs.

(e) The long-term environmental impacts of the coupling between MPs and DOCs should be examined, especially concerning the effects on human health and aquatic life. The kinetics of adsorption, the potential for bioaccumulation of organic pollutants, and the role of MPs in the transport of other contaminants (including heavy metals and persistent organic pollutants) should be further certified.

In conclusion, Yan et al. offer a valuable contribution to our understanding of the complex interactions between microplastics and dissolved organic compounds in water systems. The coupling of these two pollutants has important implications for water quality, environmental health, and water treatment processes. By shedding light on the role of microplastics in the transport and persistence of organic pollutants, this paper underscores the need for more effective pollution management strategies and the development of advanced water treatment technologies capable of addressing both microplastic and organic contamination.

2.2. *New Advances in Science and Technology for Water Purification*

Crainic and Fechete (contribution 3) compare modern and classic analytical methods for assessing the properties of slaughterhouse wastewater. Given the complex composition of slaughterhouse wastewater, which typically contains high levels of organic matter, fats, oils, and proteins, the study aims to evaluate how traditional (e.g., biochemical oxygen demand (BOD) and chemical oxygen demand (COD)) and modern techniques (e.g., advanced spectroscopic methods and sensor-based technologies) perform in analyzing its composition. The authors highlight the effectiveness of classic methods in determining basic parameters such as pH, suspended solids, and organic load, but emphasize the limitations in providing detailed insights into specific contaminants. In contrast, modern techniques offer more precise, real-time data, enabling a deeper understanding of wastewater characteristics, including the identification of specific pollutants and the variation in composition over time. This paper underscores the value of using both classic and modern methods for assessing slaughterhouse wastewater properties. The integration of these methods enables a more detailed and accurate analysis of complex contaminants, leading to improved treatment strategies. This approach could enhance the efficiency of wastewater treatment processes, particularly in industries like slaughterhouses, where wastewater characteristics are highly variable and challenging to manage.

Bule Možar (contribution 4) investigates the use of advanced oxidation processes (AOPs) as pretreatment for enhancing the biodegradation of polystyrene (PS) and polyvinyl chloride (PVC) microplastics by the fungal species Candida parapsilosis. Microplastics, particularly those made from persistent plastics like PS and PVC, are a growing environmental concern due to their stability and resistance to natural biodegradation. The AOP pretreatment with microbial biodegradation represents an innovative and effective approach to addressing microplastic pollution. This method enhances the biodegradability of otherwise resistant plastics such as PS and PVC, offering a sustainable, eco-friendly solution to plastic waste management. This paper presents a promising strategy for improving the biodegradation of PS and PVC microplastics through AOP pretreatment followed by microbial degradation using Candida parapsilosis. The findings suggest that this combined approach could be a feasible solution for addressing microplastic pollution, particularly in aquatic environments. Furthermore, the effectiveness of certain AOP techniques, such as Fenton, Photo-Fenton, and Fenton-like processes, for the degradation of microplastics including polyethylene (PE), polypropylene (PP), and polyvinyl chloride (PVC) were further evaluated in Bule Možar's report (contribution 5). The AOPs were further investigated in breaking down those persistent plastics, focusing on the comparison of Fenton, Photo-Fenton, and Fenton-like processes. These three techniques show promise for the degradation of PE, PP, and PVC microplastics. Among these, the Photo-Fenton process, enhanced by UV light, was found to be the most effective, particularly for PE and PP. Fenton-like processes, while slightly less efficient, provide a cost-effective alternative using alternative catalysts. The study calls for further research to optimize these methods for practical applications, including reductions in chemical usage, cost, and waste generation in large-scale implementations. Overall, it provides valuable insights into the potential of advanced oxidation processes for addressing microplastic pollution and offers directions for future research in this critical area.

Zhao (contribution 6) addresses a critical environmental issue—the treatment of acid mine drainage (AMD), which often contains high concentrations of toxic metals like iron (Fe) and manganese (Mn). AMD is typically highly acidic due to the oxidation of sulfide minerals in mining operations, leading to the leaching of metal contaminants into nearby water bodies. The research demonstrates that alkaline chemicals, particularly calcium hydroxide (lime), are effective in neutralizing the acidity of AMD. The treatment process successfully raises the pH of the contaminated water, leading to the precipitation of metal hydroxides, such as iron (III) hydroxide and manganese (IV) oxide. This precipitation process reduces the concentrations of dissolved metals and restores the pH to less harmful levels, making the water suitable for discharge or further treatment. The research concludes that alkaline chemical neutralization is an effective and economical method for treating AMD with high concentrations of iron and manganese. By adjusting the pH and precipitating the metals as hydroxides, this method successfully reduces metal toxicity and improves the quality of effluent water.

Chen (contribution 7) reports the use of anaerobic fermentation liquid (AFL) derived from spent mushroom substrates (SMSs) as a carbon source for nitrogen removal in wastewater treatment. Nitrogen contamination, primarily in the form of ammonium (NH_4^+), nitrate (NO_3^-), and nitrite (NO_2^-), is a major environmental concern due to its contribution to eutrophication and the deterioration of water quality. Traditional nitrogen removal methods often involve costly chemicals or energy-intensive processes, making the exploration of alternative, sustainable carbon sources highly valuable. The study demonstrates that AFL from SMSs can be used as an effective carbon source for biological nitrogen removal. The authors reveal that the AFL contains organic compounds, such as volatile fatty acids (VFAs), which can support the growth of microorganisms involved in nitrogen removal processes, such as denitrifying bacteria. The carbon-rich liquid promotes the growth of these microorganisms under anaerobic conditions, leading to efficient nitrogen removal. The AFL promotes the denitrification and ammonification step of microbial nitrogen removal processes, which demonstrates that AFL can serve as a valuable carbon source in wastewater treatment. This paper presents a novel and sustainable approach for nitrogen removal using anaerobic fermentation liquid derived from spent mushroom substrates.

Zhang (contribution 8) investigates the genotoxic effects of silver nanoclusters (AgNCs) on Daphnia magna, a widely used model organism in ecotoxicology. The study reveals that exposure to AgNCs significantly affects the genetic material of *Daphnia magna*, causing genotoxicity. The results show that AgNCs cause DNA damage in *D. magna* as indicated by increased micronuclei formation, which is a marker of chromosomal damage. This suggests that AgNCs, even at relatively low concentrations, can cause genetic alterations that may impact the organism's reproduction and overall population dynamics. A key aspect of the study is the feeding pathway. *Daphnia magna* was exposed to AgNCs indirectly by feeding on aquatic green algae that had absorbed the nanoparticles. This food chain transfer mechanism highlights the potential for nanoparticles to accumulate in lower trophic levels and be passed on to higher organisms. The paper suggests that AgNCs can be absorbed and retained by algae, and can cause genotoxicity upon ingestion by *D. magna*. The findings highlight the need for further research into the environmental and ecological risks of nanomaterials, particularly silver nanoparticles, in freshwater ecosystems. This study underscores the potential for nanoparticles to accumulate and cause genetic damage in aquatic organisms, which could have cascading effects on ecosystem health.

To ensure the safety of oil production systems and protect the underground and surface water environments of oilfields, Zhang (contribution 9) investigated the plugging issues in sand water filter pipes used in crude oil production systems. Sand plugging in filter pipes is a common challenge in oilfields, leading to decreased production efficiency, equipment damage, and increased maintenance costs. The study identifies several key materials involved in the plugging of filter pipes, including sand, crude oil components, water, and precipitates. Sand particles mixed with crude oil and water tend to accumulate and form blockages. These substances interact with water, causing scaling, fouling, and the eventual plugging of the filter system. The integration of different plugging mechanisms—physical clogging, chemical reactions, and biological fouling—into a unified framework is a significant contribution of this paper. These findings can help improve the efficiency of oilfield operations, reduce downtime, lower maintenance costs, and protect the water environment.

3. Conclusions

The two reviews (contributions 1 and 2) emphasize the importance of innovative approaches to water treatment and the need to address emerging contaminants, such as microplastics and their interaction with organic compounds, in efforts to protect aquatic ecosystems and improve wastewater management. Specifically, Wang and colleagues (contribution 1) explore the role of constructed wetlands (CWs) in wastewater treatment, with a particular emphasis on the substrates that support microbial communities and enhance contaminant removal. The review evaluates various types of substrates, such as organic, inorganic, and composite materials, and their efficiency in removing pollutants. The authors discuss the mechanisms involved, including adsorption, filtration, and bioremediation, and suggest ways to improve CW systems for more effective water purification. This approach provides an eco-friendly and cost-effective method for treating wastewater while improving water quality. In contrast, the second study by Yan and colleagues (contribution 2) examines the environmental interaction between microplastics and dissolved organic compounds (DOCs) in aquatic systems. The study investigates how microplastics, as carriers of organic pollutants, affect the distribution and behavior of DOCs, thus altering water chemistry. The authors explore the impact of microplastic contamination on aquatic ecosystems, particularly its potential to transport harmful chemicals and pathogens. This study highlights the complexity of microplastic pollution and its implications for water quality and ecosystem health.

The seven research articles provide distinct contributions to this Special Issue. Specifically, four studies (i.e., contributions 4–7) are focused on chemical (contributions 4, 5 and 6) and biological purification techniques (contribution 7). Those articles showcase a variety of innovative techniques for treating environmental pollutants, from microplastics and nitrogen to heavy metals, emphasizing sustainable and efficient approaches for pollution control and resource recovery in water treatment systems. Specifically, Bule Možar (contributions 4 and 5) explores the use of advanced oxidation processes (AOPs) as a pretreatment method to enhance the biodegradation of PS and PVC microplastics by the fungus *Candida parapsilosis* and compares the effectiveness of Fenton, Photo-Fenton, and Fenton-like processes in degrading PE, PP, and PVC microplastics. The two studies not only demonstrate that AOPs effectively break down microplastic polymers, making them more amenable to microbial degradation, providing a promising approach for mitigating microplastic pollution in aquatic environments, but also find that the Photo-Fenton process, enhanced by UV light, is particularly effective in breaking down these persistent plastics, offering a powerful method for microplastic remediation. Zhao (contribution 6) and Chen (contribution 7) examine the use of alkaline chemical neutralization to treat acid mine drainage (AMD) with high concentrations of iron and manganese, and investigate the use of anaerobic fermentation liquid from spent mushroom substrates (SMSs) as a carbon source for biological nitrogen removal. The two studies provide a sustainable and cost-effective solution for nitrogen-rich wastewater treatment and heavy metal remediation.

The other three studies focus on the monitoring measurements (contribution 3) and eco-environmental damage or security of the water environment (contributions 8 and 9). Specifically, the paper by Crainic and Fechete (contribution 3) underscores the value of using both classic and modern methods for assessing the properties of slaughterhouse wastewater. The integration of these methods enables a more detailed and accurate analysis of complex contaminants, leading to improved treatment strategies. This approach could enhance the efficiency of wastewater treatment processes, particularly in industries like slaughterhouses, where wastewater characteristics are highly variable and challenging to manage. Zhang (contribution 8) indicates that the silver nanocluster leads to significant genotoxic effects. The findings highlight the need for further research into the environmental and ecological risks of nanomaterials, particularly silver nanoparticles, in freshwater ecosystems. The long-term impacts of AgNCs on biodiversity and the potential for bioaccumulation in higher trophic levels should be monitored. Furthermore, the research conducted by Zhang (contribution 9) provides valuable insights into the plugging issues faced by sand water filter pipes in crude oil production systems and offers effective strategies for prevention and mitigation. These findings can help improve the efficiency of oilfield operations, reduce downtime, lower maintenance costs, and protect the water environment.

Author Contributions: Conceptualization, Y.Q. and K.H.; investigation, Y.Q. and K.H.; writing—original draft preparation, Y.Q. and K.H.; writing—review and editing, K.H.; All authors have read and agreed to the published version of the manuscript.

Funding: This research received no external funding.

Conflicts of Interest: The authors declare no conflicts of interest.

List of Contributions:

1. Wang, L.; Ma, L.; Wang, J.; Zhao, X.; Jing, Y.; Liu, C.; Xiao, Y.; Li, C.; Jiao, C.; Xu, M. Research Progress on the Removal of Contaminants from Wastewater by Constructed Wetland Substrate: A Review. *Water* **2024**, *16*, 1848. https://doi.org/10.3390/w16131848.
2. Yan, W.; Wang, Q.; Gao, Y.; Xu, M.; Li, H.; Zhou, Y.; Liu, C.; Xiao, Y. Coupling between Increased Amounts of Microplastics and Dissolved Organic Compounds in Water. *Water* **2023**, *15*, 4126. https://doi.org/10.3390/w15234126.
3. Crainic, R.; Fechete, R. Slaughterhouse Wastewater Properties Assessment by Modern and Classic Methods. *Water* **2024**, *16*, 2382. https://doi.org/10.3390/w16172382.
4. Bule Možar, K.; Miloloža, M.; Martinjak, V.; Ujević Bošnjak, M.; Markić, M.; Bolanča, T.; Cvetnić, M.; Kučić Grgić, D.; Ukić, Š. The Potential of AOP Pretreatment in the Biodegradation of PS and PVC Microplastics by Candida parapsilosis. *Water* **2024**, *16*, 1389. https://doi.org/10.3390/w16101389.
5. Bule Možar, K.; Miloloža, M.; Martinjak, V.; Radovanović-Perić, F.; Bafti, A.; Ujević Bošnjak, M.; Markić, M.; Bolanča, T.; Cvetnić, M.; Kučić Grgić, D.; et al. Evaluation of Fenton, Photo-Fenton and Fenton-like Processes in Degradation of PE, PP, and PVC Microplastics. *Water* **2024**, *16*, 673. https://doi.org/10.3390/w16050673.
6. Zhao, P.; Zhang, R.; Hu, M. Alkaline Chemical Neutralization to Treat Acid Mine Drainage with High Concentrations of Iron and Manganese. *Water* **2024**, *16*, 821. https://doi.org/10.3390/w16060821.
7. Chen, R.; Zhang, W.; Bi, X.; Jin, Y.; Yang, Y. Microbiological Mechanisms for Nitrogen Removal Using Anaerobic Fermentation Liquid from Spent Mushroom Substrates as a Carbon Source. *Water* **2023**, *15*, 3530. https://doi.org/10.3390/w15203530.

8. Zhang, L.; Tan, H. Genotoxic Effects on Daphnia magna Fed with Aquatic Green Algae Exposed to Silver Nanoclusters. *Water* **2023**, *15*, 3172. https://doi.org/10.3390/w15183172.
9. Zhang, W.; Liu, Q.; Chen, H.; Sheng, H.; Yan, J.; Gu, Y.; Huang, X.; Yang, B. Study on Plugging Material and Plugging Mechanism of Crude Oil Sand Water Filter Pipe. *Water* **2023**, *15*, 3714. https://doi.org/10.3390/w15213714.

References

1. Chen, J.; Xie, Y.; Sun, S.; Zhang, M.; Yan, P.; Xu, F.; Tang, L.; He, S. Efficient nitrogen removal through coupling biochar with zero-valent iron by different packing modes in bioretention system. *Environ. Res.* **2023**, *223*, 115375. [CrossRef]
2. Kumar, V.; Parihar, R.D.; Sharma, A.; Bakshi, P.; Singh Sidhu, G.P.; Bali, A.S.; Karaouzas, I.; Bhardwaj, R.; Thukral, A.K.; Gyasi-Agyei, Y.; et al. Global evaluation of heavy metal content in surface water bodies: A meta-analysis using heavy metal pollution indices and multivariate statistical analyses. *Chemosphere* **2019**, *236*, 124364. [CrossRef] [PubMed]
3. Campos, L.C. A Review of the Most Concerning Chemical Contaminants in Drinking Water for Human Health. *Sustainability* **2024**, *16*, 7107. [CrossRef]
4. Singh, P.K.; Kumar, U.; Kumar, I.; Dwivedi, A.; Singh, P.; Mishra, S.; Seth, C.S.; Sharma, R.K. Critical review on toxic contaminants in surface water ecosystem: Sources, monitoring, and its impact on human health. *Environ. Sci. Pollut. Res.* **2024**, *31*, 56428–56462. [CrossRef] [PubMed]
5. Nehru, R.; Chen, C.W.; Dong, C.D. A review of smart electrochemical devices for pesticide detection in agricultural food and runoff contaminants. *Sci. Total Environ.* **2024**, *935*, 173360. [CrossRef]
6. Ma, Y.; Gu, X.; Zhang, Y.; Yan, P.; Zhang, M.; Sun, S.; Ren, T.; Tang, L.; He, S. Unveiling the microplastic perturbation on surface flow constructed wetlands with macrophytes of different life forms: Responses of nitrogen removal and sensory quality. *J. Hazard. Mater.* **2024**, *477*, 135283. [CrossRef] [PubMed]
7. Sakkaravarthy, S.; Jano, N.A.; Vijayakumar, A. *Overcoming Challenges in Traditional Waste Water Treatment Through AI-Driven Innovation*; Springer: Cham, Switzerland, 2024. [CrossRef]
8. Sangamnere, R.; Misra, T.; Bherwani, H.; Kapley, A.; Kumar, R. A critical review of conventional and emerging wastewater treatment technologies. *Sustain. Water Resour. Manag.* **2023**, *9*, 58. [CrossRef]
9. Adesina, O.B.; William, C.; Oke, E.I. Evolution in Water Treatment: Exploring Traditional Self-Purification Methods and Emerging Technologies for Drinking Water and Wastewater Treatment: A Review. *World News Nat. Sci.* **2024**, *53*, 169–185.
10. Qi, Y.; Ge, B.; Zhang, Y.; Jiang, B.; Wang, C.; Akram, M.; Xu, X. Three-dimensional porous graphene-like biochar derived from Enteromorpha as a persulfate activator for sulfamethoxazole degradation: Role of graphitic N and radicals transformation. *J. Hazard. Mater.* **2020**, *399*, 123039. [CrossRef]
11. Peng, Y.; Gu, X.; Zhang, M.; Yan, P.; Sun, S.; He, S. Simultaneously enhanced autotrophic–heterotrophic denitrification in iron-based ecological floating bed by plant biomass: Metagenomics insights into microbial communities, functional genes and nitrogen metabolic pathways. *Water Res.* **2024**, *248*, 120868. [CrossRef] [PubMed]
12. Lu, X.; Mu, C.; Liu, Y.; Wu, L.; Tong, Z.; Huang, K. Recent advances in solar-driven interfacial evaporation coupling systems: Energy conversion, water purification, and seawater resource extraction. *Nano Energy* **2024**, *120*, 109180. [CrossRef]
13. Liberti, D.; Pinheiro, F.; Simões, B.; Varela, J.; Barreira, L. Beyond Bioremediation: The Untapped Potential of Microalgae in Wastewater Treatment. *Water* **2024**, *16*, 2710. [CrossRef]

water

MDPI

Review

Research Progress on the Removal of Contaminants from Wastewater by Constructed Wetland Substrate: A Review

Liyan Wang [1], Leihui Ma [1], Junke Wang [1], Xia Zhao [1], Yushu Jing [1], Changqing Liu [2], Yihua Xiao [2], Cang Li [2], Chen Jiao [2] and Mengchen Xu [2,3,*]

[1] Qingdao Zhangcun River Water Co., Ltd., 908-1# Tongan Road, Qingdao 266000, China; 17860392185@163.com (L.W.); w846292989@163.com (L.M.); e5981919@126.com (J.W.); 19553219596@126.com (X.Z.); jysqd@163.com (Y.J.)

[2] School of Environmental and Municipal Engineering, Qingdao University of Technology, Qingdao 266033, China; lcqlfy@163.com (C.L.); yihua.xiao@qut.edu.cn (Y.X.); l498416156@126.com (C.L.); jc5499165@126.com (C.J.)

[3] School of Environmental Science and Engineering, China-America CRC for Environment & Health, Shandong Province, Shandong University, 72# Jimo Binhai Road, Qingdao 266237, China

* Correspondence: xumengchen@qut.edu.cn

Abstract: Constructed wetlands (CWs) primarily achieve efficient wastewater purification through synergistic interactions among substrates, plants, and microorganisms. Serving as the structural foundation of the entire wetland system, substrates not only provide a growth medium for plants, but also serve as adhesive carriers for microorganisms and habitats for animal activities. Research on substrates has attracted considerable attention; however, in practical engineering applications, the selection of substrates often depend on personal experience, which may lead to significant gaps in the effectiveness of wetland systems in treating different characteristic contaminants. Therefore, it is of great significance to investigate the influence of substrates on the removal of contaminants in sewage and identify substrate materials with good physical and chemical properties to optimize the design and operation of CWs-based sewage-treatment systems and improve their purification efficiency. In this review, bibliometric analysis was conducted to using the Web of Science database and VOSviewer_1.6.20 software to assess the progress of research on CWs. This article provides a comprehensive overview of substrate types and characteristics based on recent research advancements in the field. Additionally, it discusses removal methods and the influence of factors related to conventional contaminants (COD, nitrogen, and phosphorus), heavy metals (HMs), fluorinated compounds, pharmaceuticals, personal care products (PPCPs), and microplastics. A thorough evaluation was conducted on the economic costs of various substrates and their ability to remove major contaminants from water bodies, providing a reference for the further development of wetland technology.

Keywords: constructed wetlands; substrates; emerging contaminants; wastewater treatment function

1. Introduction

CWs are ecosystems built by artificial simulation of natural processes, usually relying on the synergistic interaction of substrates, plants, and microorganisms to enhance wastewater purification through physical, chemical, and biological processes [1,2]. The physical effects primarily encompass filtration and retention, which heavily rely on collaboration between the substrate layer and plants to effectively intercept and adsorb contaminants [3]. The chemical reactions involved in this study encompass adsorption, chemical precipitation, redox reactions, and ion exchange. These processes are primarily governed by the substrate's chemical properties [4]. Biological reactions include plant metabolism and the growth activities of various microorganisms. Microorganisms are attached to the substrate and plant roots. Various microorganisms remove contaminants through the processes of absorption, metabolism, and transformation [5]. Biological reactions are the most important step in the removal of contaminants in CWs [6].

The findings of various studies have demonstrated that CWs possess the capability to effectively reduce the concentration of contaminants in domestic sewage. Moreover, CWs have also exhibited their potential in treating industrial wastewater [7], mine effluents [8], municipal effluents [9], hospital wastewater [10], agricultural wastewater [11], etc. By the end of 2023, more than 11,000 articles about CWs had been published on Web of Science. In order to better understand the current development status of CWs, "constructed wetland" was used in a keyword search of the literature from the last five years (2019–2023). After screening, a total of 5140 publications meeting the requirements were selected. The VOSviewer program was used for analysis, and the minimum frequency of occurrence was set to 100. A total of 81 keywords were included in the visualization. The outcomes of the analysis on highlighted words are presented in Figure 1. The figure exhibits four distinct color clusters, namely red, yellow, blue, and green. Words in the red cluster are mainly related to the removal of nitrogen and phosphorus, such as nitrogen, phosphorus and nutrient. Words in the yellow cluster are closely related to plants, such as phytoremediation, macrophytes, and plants. Words in the blue cluster are closely related to the direction of future research on CWs, such as personal-care products, pharmaceuticals and fate. The green cluster mainly relates to the mechanism of the removal of contaminants, such as denitrification and nitrification.

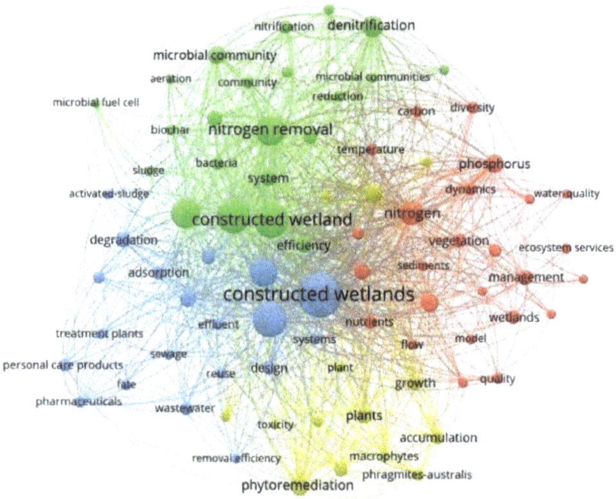

Figure 1. Visualization of keywords in the 2019–2023 CW research.

Table 1 shows the keywords with the highest frequency in CWs. It can be seen that the current research on CWs is still focused on traditional wastewater treatment, especially the removal of nitrogen and phosphorus contaminants (sort 5, 8 and 12). However, the increasingly intricate composition of contaminants in wastewater and the growing public demand for high water quality, particularly with regards to the emergence of various EPs, present novel challenges to CWs [12]. Studies have shown that the traditional substrate (soil, gravel, sand) mainly plays a supporting role for plants, but its effect on the removal of contaminants is often unsatisfactory [13]. Therefore, in the design of substrate for CWs, reasonable configurations should be selected to improve the efficiency of treatment of contaminants.

Table 1. 2019–2023 CWs keywords prominence results table.

Sort	Keywords	Frequency
1	Constructed wetlands	1546
2	Constructed wetland	1029
3	Removal	1019
4	Performance	993
5	Nitrogen removal	796
6	Waste-water treatment	772
7	Waste-water	712
8	Nitrogen	530
9	Denitrification	446
10	Phytoremediation	350
11	Nutrient removal	333
12	Phosphorus	331

The substrates form the structural framework of the entire wetland, serving as a fundamental substrate for plant growth, an attachment surface for microorganisms, and a habitat for animal activities [14,15]. The physical properties of the substrate (particle size, porosity, specific surface area, and redox potential) will determine the effectiveness of treatment by CWs. For example, a substrate with increased specific surface area and porosity can facilitate biofilm formation and attachment and enhance microorganism–contaminant interactions at multiple contact points. According to the China Technical Specification of Constructed Wetlands for Wastewater Treatment (HJ 2005–2010) (MEPC, 2010), the selection of substrate should meet the following requirements: (i) good mechanical strength, high specific surface area and high stability; (ii) appropriate size in accordance with local sampling; (iii) when the content of nitrogen and phosphorus in the treated sewage is high, a functional substrate should be used to improve removal; (iv) for vertical subsurface flow (VSSF) CWs, the porosity of the substrates should be between 35–40%. Although research on substrates has attracted considerable attention, in practical engineering applications, the selection of substrate often relies on personal experience, which may lead to significant gaps in the effectiveness of wetland systems in treating different characteristic contaminants. Therefore, it is crucial to investigate the influence of substrate on the removal of contaminants in sewage and identify substrate materials with good physical and chemical properties to optimize the design and operation of CW-based sewage-treatment systems and improve their purification efficiency. This article summarizes the types and characteristics of substrates in CWs based on recent research achievements in the field of wetlands. It further discusses the removal mechanisms associated with substrates and the influence of factors specific to conventional contaminants (COD, nitrogen, and phosphorus), HMs, fluorinated compounds, PPCPs, and microplastics. A comprehensive evaluation was conducted on the economic costs of various substrates and their ability to remove major contaminants from water bodies, providing a reference for the further development of wetlands technology.

2. Typical/New Substrates for CWs

According to the source of the materials, the substrates of CWs can be divided into natural substrates, by-products and artificial synthetic substrates. Table 2 lists the representative substrates used in CWs at present. The physical and chemical properties of different substrates exhibit significant variations, thereby resulting in diverse impacts on CWs [16]. Consequently, during the operation of CWs, it is crucial to select appropriate substrates or combinations thereof based on the specific contaminant-removal requirements, as this may enhance the overall performance of CW systems.

2.1. Natural Substrates

The most common natural substrates in CWs are gravel, sand, volcanics and zeolite. In general, natural substrates are made from natural mineral materials and have the advantages of high strength, a wide range of sources, and low price. However, the permeability coefficient and specific surface area tend to be low, resulting in poor performance in contaminant removal. Therefore, the natural substrates are often used as a supporting part of the filter bed to reduce the design cost of CWs [17]. There are also studies comparing natural substrates with different substrates in order to find a material that is good at removing particular contaminants. The study conducted by Huo et al. [18] involved the construction of four CWs utilizing different substrates (gravel, biochar, zeolite, and pyrite). The results indicate that the CWs with gravel as a substrate exhibited the least efficient nitrogen removal. Microbiological analysis revealed a low abundance of nitrifying and denitrifying microorganisms in the gravel-based CW system.

2.2. *By-Products*

2.2.1. Agricultural By-Products

The term "agricultural by-products" refers to the waste generated during agricultural production, which mainly includes various crop residues, livestock manure, and processing by-products. In 2023, China's grain output reached 695 million tons, which will undoubtedly produce a large number of agricultural by-products. China has emerged as the leading country in terms of agricultural by-products, displacing up to four billion tons of waste annually. However, improper utilization of these by-products can pose significant environmental risks [19]. Currently, it is increasingly common to use agricultural by-products as the substrate of CWs, which not only uses this resource, but also reduces the cost of CWs and improves their effectiveness in treatment [20,21].

2.2.2. Industrial By-Products

Industrial by-products are mainly substances that are inevitably generated during the production of target products in industrial processes. At present, the global production of industrial by-products can reach billions of tons per year [22]. The inadequate treatment of industrial by-products may lead to severe environmental issues if it is not addressed promptly. Consequently, the exploration of resource utilization for industrial by-products has emerged as a prominent research focus. Common industrial by-products used in CWs include fly ash, steel slag, iron filings, and construction solid waste. Some by-products have good physical and chemical properties, high porosity, and high specific surface area. Due to differences in the raw materials and treatment processes, different industrial by-products show significantly different performance in treating contaminants. Furthermore, it is important to note that industrial by-products may contain a substantial amount of toxic and hazardous substances, thereby posing potential environmental risks [23]. In practical applications, it is advisable to prioritize the use of harmless materials. When selecting by-products or other materials that may potentially impact the environment, appropriate modifications should be implemented to ensure their harmlessness [24,25].

2.3. *Artificial Synthetic Substrates*

Artificial synthetic substrates are new materials formed by calcination, carbonization, and modification, either individually or in combination with various materials. The common substrates in CWs include activated carbon, biochar, ceramsite, and modified materials. The physical and chemical properties of these materials are stability and high porosity. However, due to the influence of the preparation process and raw materials, they exhibit different levels pf performance in treating different contaminants. In addition, artificial synthetic substrates also have the problem of high cost, which have raised serious obstacles to their promotion and use. At present, the use of artificial synthetic substrates is still mainly a subject of laboratory research. Developing a low-cost and high-efficiency filler has become a hot topic for research [26,27].

Table 2. Classification of representative CWs substrates.

Category	Substrate	Bases	Sources	Characteristic	Reference
Natural substrate	Gravel	SiO_2 and Al_2O_3	Rock or mineral detritus	① Low cost ② Poor removal of contaminants	[28]
Natural substrate	Sand	SiO_2	Tiny stone particles	① Availability and low cost ② Easily clog CWs	[29]
Natural substrate	Volcanics	SiO_2, Fe_2O_3, CaO and Al_2O_3	Cooling magma after a volcanic eruption	① High porosity and specific surface area ② Poor contaminant removal	[30]
Natural substrate	Zeolite	SiO_2 and Al_2O_3	Mainly derived from alkaline groundwater and volcanic rocks	① High ion exchange, low cost, can be regenerated ② Low adsorption rate, easy to be affected by the environment	[31,32]
By-product	Fly ash	Fe_2O_3, Al_2O_3, SiO_2 and CaO	Waste from burning coal	① Large specific surface area, good adsorption performance ② Can easily cause secondary pollution in the environment	[33]
By-product	Steel slag	CaO, SiO_2 and Fe_2O_3	From electric furnaces and refining furnaces during steelmaking	① High phosphorus-removal performance ② Strong alkalinity, which makes it difficult for microorganisms and plants to survive	[34]
By-product	Construction solid waste	-	Waste generated during demolition, construction and maintenance in the construction industry	① Low cost, high availability ② Poor reusability	[35]
By-product	Sludge	Organics, SiO_2 and Al_2O_3	Waste from drinking water treatment	① Low cost, high capacity for P and HMs adsorption, easy to obtain ② Very sensitive to the environment, might leach toxic substances	[14]
By-product	Iron scraps	Fe	Waste from steel mills and machinery-processing plants	① Low cost, large specific surface area	[36,37]
By-product	Woodchip	Cellulose, hemicellulose, lignin	Waste from wood processing	① Low cost, excellent N removal ② May cause secondary pollution	[38]
By-product	Crushed glass	Na_2SiO_3, CaO and SiO_2	Waste from glass production and use	① Wide range of sources, reusable, provides space for microbes to attach	[39]
Artificial synthetic substrate	Ceramsite	SiO_2, Al_2O_3 and CaO	Made by roasting raw materials	① Good adsorption performance, high strength ② High energy consumption	[40]
Artificial synthetic substrate	Activated carbon	Porous carbon	Made by carbonization of raw materials	① Moderate adsorption effect ② High cost	[41]
Artificial synthetic substrate	Biochar	Porous carbon	Made by biomass pyrolysis	① Excellent contaminant adsorption, stable ② Difficulty in preparation	[42]
Artificial synthetic substrate	Modified substrates	-	Further processing of materials	① Excellent contaminant adsorption ② High cost and high energy cost	[43]

3. Functions of Substrates Removal of Contaminants

3.1. Substrates Removes Conventional Contaminants

3.1.1. Organic

The removal of organic matter in CWs mainly primarily relies on biodegradation by microorganisms [44]. As the site of microorganism metabolism, the substrates play an important role in the growth of microorganisms. In addition, the substrate can directly remove contaminants through adsorption and retention, which heavily rely on collaboration between the substrate layer and plants [26].

The physicochemical properties of the substrates exert a significant influence on the removal of organic compounds, with porosity being identified as one of the pivotal factors. Several studies have demonstrated that during continuous operation of CWs, there is an observed increase in insoluble organic matter, leading to a gradual reduction in substrate porosity and potentially resulting in clogging [45]. Since the wetland system mainly relies on microbial oxidative decomposition, anaerobic digestion, and other life activities to purify organic matter, clogging can result in extensive anoxia within the CWs as a whole. This condition diminishes the metabolic capacity of microorganisms and consequently lowers the efficiency of organic-matter removal.

In order to increase the removal of organic matter, substrates with high porosity such as zeolite and ceramsite can be selected. Wang et al. [46] utilized sewage sludge as a raw material to prepare new ceramic particles. A VSSF-CW was constructed with this new type of ceramic particle as the substrate. The COD-removal rate reached more than 70%, while the effluent quality reached the IV water standard. The high degradation rate of COD may be attributed to the substantial specific surface area and porosity of the novel ceramic particles, which facilitate enhanced contact between microorganisms adhered to the substrate surface and organic matter. Wang et al. [47] modified corn-straw biochar by alkali treatment. The results demonstrated that the specific surface area and porosity of the modified biochar were higher than they were before modification and that the effectiveness of treatment of other contaminants was also significantly improved.

3.1.2. Nitrogen

Nitrogen removal in CWs mainly depends on the processes of ammonification, nitrification, and denitrification by microorganisms. Organic nitrogen is first converted into NH_4^+-N through ammonification, which is followed by aerobic transformation into NO_2^--N and NO_3^--N by nitrifying bacteria. Finally, denitrifying bacteria convert NO_3^--N into N_2 under anaerobic conditions, so as to achieve nitrogen purification. The selection of carbon-rich fillers such as biochar, straw, and wood chips can improve the nitrogen-removal effect. Wang et al. [48] utilized corn stalks to prepare activated carbon for urban tailwater treatment, and this material exhibited an impressive nitrogen-removal efficiency, exceeding 85% in urban tailwater treatment. They concluded that using carbon-rich materials as the substrate of CWs can augment the content of organic matter and the C/N ratio, thus promoting the denitrification reaction. Furthermore, in addition to supplementation of carbon sources, the limitation on denitrification caused by low C/N in wetland influent can also be solved by oxygenation reduction reactions induced by inorganic reductive fillers. Common materials include nano zero-valent iron (NZVI) [49], pyrite [50], and sulfur [51]. According to the reaction mechanism of reduction, these materials can serve as electron donors for NO_3^--N to achieve nitrogen removal in a process referred to as autotrophic denitrification [37]. Chu et al. [52] established eight sets of VSSF-CWs supplemented with pyrite. The results showed that the denitrification capacity of the pyrite-addition device was improved and that the TN removal rate was more than 27.05% higher compared to that in the control group. DO content also has an important impact on wetland systems. Insufficient DO content hampers the process of nitrification, thereby impeding subsequent denitrification and resulting in limited nitrogen-removal efficiency [53]. Some researchers have found that selecting substrates with higher porosity can not only enhance microbial attachment, but also help increase the DO content [17]. Zhou et al. [54] studied a new type of CWs that use biochar as the substrate. The high porosity of biochar promotes efficient oxygen diffusion within CWs, thereby enhancing nitrogen-removal performance and resulting in a TN-removal rate of 39%. In addition, apart from direct nitrogen removal via physical adsorption, the substrate can also remove NH_3-N through ion-exchange pathways. Zeolite is a common material with high ion-exchange capacity. The zeolite lattice is composed of SiO_4 and AlO_4 tetrahedra, wherein the presence of aluminum imparts a negative charge to the framework. To maintain electrical neutrality, cations are required for charge balancing [55]. These cations (generally Na^+ and K^+.) are easily exchanged with NH_4^+ to remove ammonia nitrogen ions. Alshameri et al. [56] modified Yemeni zeolite by alkali treatment to allow it to adsorb ammonia nitrogen. The results showed that the removal of ammonia nitrogen by zeolite mainly depended on ion exchange, and the rate of removal of nitrogen by modified zeolite was as high as 99%.

3.1.3. Phosphorus

The removal of phosphorus in CWs mainly depends on microbial fixation, plant absorption, soil increment, and substrate adsorption [30,57,58]. Research has shown that the substrate can remove 36.2–87.5% of the phosphorus present in the water [59]. Phosphorus adsorption by the substrate encompasses physical adsorption and chemical precipitation processes, with chemical precipitation playing a significant role [35]. Phosphorus removal by chemical precipitation is mainly a process in which some substrates rich in Ca, Fe, and Al can combine with phosphate to form an insoluble phosphate precipitate (Figure 2a). The efficiency of chemical phosphorus removal is pH-dependent due to the existence of phosphate in various forms at different pH levels. When the sewage environment is weakly acidic or neutral, the predominant inorganic phosphate groups are $HPO_4{}^{2-}$ and $H_2PO_4{}^{-}$. Conversely, in a weakly basic or alkaline environment, the main inorganic phosphate groups are $HPO_4{}^{2-}$ and $PO_4{}^{3-}$. Research has demonstrated that metal ions exhibit varying phosphorus-removal efficiencies at different pH levels. Within the pH range of 4.5–7.21, Fe-Al-based phosphorus removal dominates, with a maximum phosphorus removal capacity of 17.88–21.63 mg/g. On the other hand, when the pH ranges from 5.42 to 9.83, Ca becomes the primary method for phosphorus removal, with an effective capacity ranging from 20.85 to 21.43 mg/g [60]. The selection of Ca, Al, and Fe-rich substrate materials will increase the phosphorus-adsorption capacity. The presence of metal ions in certain by-products and artificial synthetic materials makes them suitable as substrates for phosphorus removal. Jiang et al. [61] compared the phosphate-removal properties of quartz sand, anthracite, shale, and ceramics. The results showed that the anthracite and bio-ceramics had better removal effects on TP, at 85.87 and 81.44 mg/kg, respectively. The shale and quartz sand had weaker removal properties, at 59.65 and 55.98 mg/kg, respectively. The substrate can also be modified to improve phosphorus removal. Guaya et al. [62] modified zeolite by adding hydrated alumina oxide (HAlO) and showed that the adsorption properties of zeolite increased from 0.7 mg/g to 7.0 mg/g. The modification does not have a significant impact on the substrate's original chemical properties. Stocker et al. [63] modified zeolite with Ca and showed through joint experiments with NH_3-N and $PO_4{}^{3-}$ that the modified zeolite did not have a significant impact on the removal of ammonia nitrogen but did remove phosphate. The purification efficiency of CWs with regard to phosphorus is not only affected by the physical properties of the substrate itself but is also related to external factors such as HRT, temperature, pH, phosphorus concentration, and DO (Figure 2b). The use of a suitable operation mode is the key to achieving efficient phosphorus removal in CWs. Moreover, it is crucial to acknowledge that phosphorus removal in CWs can be achieved only through the harvesting of plants or the removal of saturated substrates. Failure to replace the substrate promptly will result in the release of previously adsorbed phosphorus back into the water body.

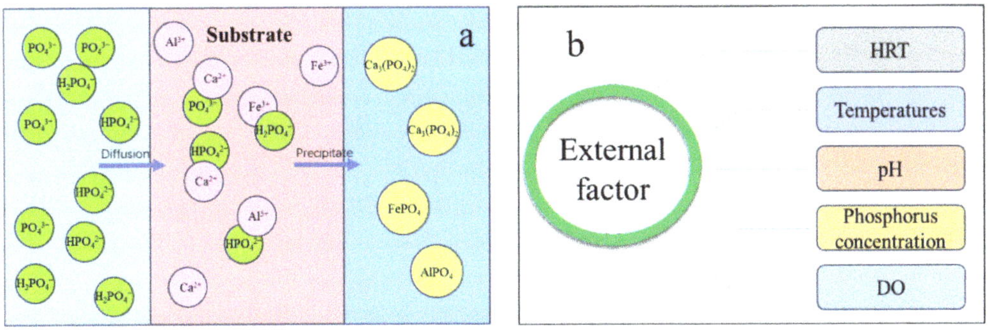

Figure 2. (**a**) Mechanism of chemical precipitation; (**b**) Influence of external factors on phosphorus removal by the substrate.

3.2. HMs Removal

Due to economic development and the acceleration of industrialization, the emission of HMs has soared. However, it is currently observed that numerous countries, particularly developing countries, exhibit deficiencies in their wastewater treatment. This inadequacy may result in the direct discharge of substantial quantities of HMs into natural water bodies without any form of treatment, thereby posing a considerable threat to human health [64,65]. In recent decades, there has been growing interest in the use of CWs to remove HMs. The methods of HMs removal by CW mainly include sedimentation, adsorption, chemical precipitation, ion exchange, flocculation, microbial interaction, and plant absorption [66,67]. Insoluble HMs in water can be removed by sedimentation and retention, while ion exchange and chemical precipitation can effectively transform soluble HMs into insoluble or insoluble compounds [68]. The mechanism of HMs ion exchange is analogous to that of NH_3-N ion-exchange removal. However, it should be noted that the performance of fillers for the removal of different ions varies due to disparities in ionic charge, water solubility, pK_H, and ionic strength [69]. Lee et al. [70] explored adsorption by different modified biochars of Cd, Pb, and Zn. The results showed that the adsorption capacity followed the order $Pb^{2+} > Cd^{2+} > Zn^{2+}$. This can be attributed to the smaller hydration radius and relatively lower pK_H value of Pb^{2+}, which facilitates its easier adsorption. Moreover, chemical precipitation is also an important way to remove HMs. Tang et al. [71] prepared iron-manganese oxide-modified biochar (FM-BC) for the adsorption of Pb (II) in sewage. The results revealed that the removal of Pb (II) mainly depended on the co-precipitation and complexation of Fe and Mn. Common methods of HM removal by precipitation also include hydroxide precipitation, sulfide precipitation, and carbonate precipitation [68]. Due to the complexity of contaminant-removal mechanisms in CWs, the substrate often serves multiple roles in the removal of HMs. In order to improve the effectiveness of HMs removal from wastewater, Kong et al. [27] prepared a Fe-Al bimetallic oxide modified zeolite to remove Cr (VI) from wastewater. When the initial concentration of Cr (VI) is 20 mg/L, the removal rate can reach 84.96% due to the oxidation–reduction reaction and co-precipitation between Cr (VI) and Fe-Al.

3.3. Fluoride Compounds Removal

Fluorine, a naturally occurring element, readily forms fluoride through chemical bonding with other elements, serving as a vital constituent for the development of teeth and bones. However, excessive exposure to fluoride can lead to dental fluorosis and even skeletal deformities [72,73]. CWs mainly rely on adsorption and coprecipitation of substrate to remove F^-. Yao et al. [74] used limestone as the substrate to increase the calcium content in wetlands. XRD showed that the formation of CaF_2, $Ca_5(PO_4)_3$, and $Ca_3(PO_4)_2$ precipitates mainly depended on adsorption and co-precipitation. It is worth noting that the researchers found that the fluoride-removal effect of CWs with limestone as the substrate was not significantly different from that of the control group. This suggests that calcium-based wetlands do not show improved fluoride-removal performance. A special organic fluoride named perfluoroalkyl and polyfluoroalkyl substances (PFAS) have received extensive attention due to their toxicity and capacity for bioaccumulation [75]. CWs, synergistically with microorganisms, plants and substrates, may be able to promote their removal from water bodies. Ma et al. [76] added iron mineral to VSSF-CWs to remove perfluorooctane sulfonic acid (PFOS) and perfluorooctanoic acid (PFOA) from water. The results showed that the dosage of iron mineral increased the concentration of microorganisms capable of removing PFAS (e.g., *Burkholderia* and *Pseudomonas*). Compared with the control group, the effluent PFOS and PFOA contents decreased by 35.0% and 36.8%.

3.4. PPCPs Removal

PPCPs come from various drugs, disinfectants, fragrances, and cosmetics [77]. Upon entering the aquatic environment, these substances can potentially diminish the survival rate of aquatic organisms. More importantly, their accumulation in such organisms poses a significant risk to human health through consumption [78]. Unfortunately, current wastewater treatment plants (WWTPs) mainly focus on the removal of conventional contaminants, often neglecting the elimination of PPCPs. As a result, some PPCPs manage to survive in traditional WWTPs and are subsequently discharged into natural water bodies [79]. CWs are considered an effective way to deal with PPCPs due to their low cost and environmental friendliness, especially as a method for removing antibiotics and antibiotic-resistance genes (ARGs). Chen et al. [80] established four CWs based on zeolite, resulting in impressive removal efficiencies of 87.4% for antibiotics and 87.8% for ARGs. The study revealed that microbial degradation emerged as the predominant factor, followed by substrate adsorption and plant absorption. Liu et al. [81] used zeolite as the substrate to remove sulfamethoxazole (SMX) and ofloxacin (OFLO). They found that the mechanism of removal of SMX and OFLO by zeolite was actually an adsorption–degradation process. When the zeolite comes into contact with the antibiotic, it rapidly undergoes adsorption onto the surface of the zeolite, which is followed by biodegradation facilitated by microorganisms attached to the zeolite. Subsequently, the excess adsorption sites on the zeolite surface will continue to adsorb antibiotics, reducing the concentration of antibiotics. Use of a suitable adsorbent can improve the removal of PPCPs. Rodriguez et al. [82] explored the use of powdered activated carbon in combination with ultrafiltration to remove five PPCPs (1-H-benzotriazole, DEET, chlorophene, 3-methylindole, and nortriptyline-HCl) and found that adsorbents can significantly promote the removal of compounds with poor hydrophobicity. Pore size plays an important role in removing PPCPs, and powdered activated carbon with high microporosity can better adsorb PPCP and organic matter.

3.5. Microplastics Removal

Microplastics are plastic particles with a diameter of less than 5 mm. The emission of microplastics not only leads to ecological pollution, but also inflicts irreversible harm upon the health of aquatic organisms [83,84]. Microplastics can be removed by adsorption, filtration, sedimentation, flocculation, and biodegradation [85]. CWs have been reported as a good new way to remove microplastics [86]. The mechanism of removal of microplastics by CWs mainly relies on physical retention. The substrate, serving as the carrier within CWs, plays an important role in the removal of microplastics. In addition, the substrate can also be used to remove microplastics by adsorption and chemical precipitation. Chen et al. [87] studied the removal of microplastics by CWs, and the results showed that most microplastics were removed through the substrate layer of 0–20 cm at the water inlet. Selecting a substrate with a larger pore size could enable the microplastics to be transported farther and improve the removal of microplastics. In addition, since microplastics generally carry a negative charge, cations can be used to capture and remove them. Shen et al. [88] used cationic surfactants to modify an aluminosilicate filter medium, and the capture rate of microplastics reached more than 96%, much higher than that of the rapid sand filter (63%). The utilization of CWs for the removal of microplastics has emerged as a prominent area of research; however, due to an incomplete understanding of the underlying removal mechanisms, current experimental investigations on microplastic removal using CWs are primarily conducted within laboratory settings.

3.6. Chapter Summary

The above content summarizes the role of CW substrates in the removal of various contaminants. Generally speaking, substrates can directly participate in the removal of contaminants through filtration and interception, adsorption, chemical precision, redox reactions, and ion exchange. Additionally, they serve as the foundation for CWs by facilitating the growth and reproduction of microorganisms and plants. Therefore, selecting the appropriate substrate is crucial for wetlands. Table 3 summarizes 14 representative substrate materials and evaluates their ability to purify water, providing a reference for the selection of high-quality substrates.

Table 3. The removal effect of different substrates on contaminants.

Contaminants Type	COD	N	P	Heavy Metals	Fluoride	PPCPs	Microplastics
Gravel	+	+	++	++	-	+	+
Sand	+	+	+	+	-	++	++
Volcanics	+	++	+	+	++	-	-
Zeolite	++	+++	+	++	++	-	++
Fly ash	+	++	+++	-	+	-	-
Steel slag	++	++	++	-	-	-	-
Construction solid waste	+	++	+++	-	-	-	-
Sludge	++	+	+++	-	-	-	-
Iron scraps	+	+++	+++	+	-	-	-
Woodchip	+++	++	+++	-		-	-
Crushed glass	-	-	-	-	-	++	-
Ceramsite	+++	+	+++	++	++	+	-
Activated carbon	+++	+++	+++	+++	+++	+++	+++
Biochar	+++	+++	+++	+++	++	++	+++

Note: +, ++, +++ respectively mean average, good, very good, - means poor effect or not yet studied.

4. Problems and Prospects

4.1. Clogging

At present, the clogging problem has become a key factor restricting the long-term operation of CWs. The main causes of substrate clogging include deposition of various solids, precipitation, and microbial and plant growth [89]. Substrate clogging decreases the permeability coefficient and porosity of the substrate, which reduces the treatment performance and service life of CWs [29]. Choosing the right substrate can effectively reduce clogging. Wang et al. [90] used NaCl as a tracer to test the changes in electrical conductivity (EC) of nine substrates (quartz sand, zeolite, coarse sand, gravel, two types of biochar, and three types of ceramsite). The findings suggest that zeolite has minimal impact on the electrical conductivity (EC) of wastewater, whereas the utilization of artificial synthetic substrates like sludge biochar can significantly elevate EC levels in water, thereby increasing the susceptibility of wetlands to clogging. In addition, the adsorption capacity of the substrate is limited [91]. With increased substrate adsorption, the degree of clogging of CWs will continue to rise. Furthermore, once the adsorption reaches saturation, contaminants will escape from the adsorbent and enter the water body [92]. Therefore, the substrate needs to be replaced before it becomes clogged.

4.2. Substrate Selection and Collocation

Currently, problem of pollution associated with sewage has become increasingly complex, and relying solely on a single substrate often makes it difficult to completely remove contaminants. Developing substrates with different contaminant-removal characteristics and combining them can achieve better contaminant removal through the synergistic effect between substrate layers. Composite substrates can extend the life cycle and improve the removal of emerging pollutants. Xu et al. investigated the synergistic effect of a mixed substrate comprising manganese ore and activated alumina on enhancing the efficiency of contaminant removal. The combined approach demonstrated remarkable performance, achieving 89.4% COD removal and 98.1% TP removal. Under different types of flow direction, the optimal layout of substrates is significantly different [16]. In free water-surface-flow (FWS) CWs, the major function of substrates is to support the plants. The hydraulic permeability of substrates should be the focus in the design of vertical subsurface (VSSF) CWs. In recent research, a VSSF–horizontal subsurface flow (HSSF)-FWS hybrid system was constructed to treat domestic wastewater and showed better treatment efficiency [93–95]. In hybrid CWs, various substrates were applied at different stages to obtain better treatment efficiency.

However, there is currently no standardized approach for selecting the substrate type and determining the optimal mixing ratio of CWs, while the removal mechanism remains a subject of ongoing research. It is certain that the combination of substrates has become an important method by which to improve the treatment performance of contaminants and improve the service life of wetlands [96].

4.3. Combining CW Technology with Other Processes

CWs are widely used due to their advantages of cost effectiveness and ease of maintenance and management. However, major parts of the wetland do not have access to sufficient electron donors, especially oxygen, which leads to inefficient contaminant removal [97]. Combining CWs with other technologies can help ameliorate the lack of electron donors and thus improve contaminant removal. Technologies commonly combined with CWs include photocatalysis [98], microbial fuel cells (MFC) [99], membrane bioreactors (MBR) [100], etc. However, due to differences in water quality and the types and quantities of contaminants in different regions, there is still no scientific theoretical basis and standard for the research on the combination of different CWs with other technologies, which limits the promotion and application of such systems. Exploring the operating mechanism of composite systems and selecting combinations with good processing effects and strong adaptability can improve the overall processing capacity of the wetland and reduce the land area required.

5. Conclusions

This article reviews the research achievements in the field of wetlands in recent years, categorizes and summarizes the types of substrates, and elaborates on the removal methods and the influence of contaminant-related factors (conventional contaminants and EPs) in CWs. The physical properties of various substrates were comprehensively evaluated, along with their efficacy in contaminant removal, providing valuable insights for the further advancement of CWs. In summary, through a literature review of substrates, it can be concluded that:

1. CWs have been widely used in wastewater treatment. At present, there are more than 11,000 articles related to CWs in the Web of Science database. The term "constructed wetland" was used as the key word to search the database. In the past five years (2019–2023), a total of 5140 relevant publications were found after deduplication. Afterwards, VOSviewer software was used to visually analyze these 5140 articles, and the results show that the current research on CWs is still mainly focused on the removal of traditional contaminants (e.g., nitrogen and phosphorus).
2. According to the source of the substrates, they can be divided into natural substrates, by-products (agricultural by-products and industrial by-products) and artificially synthesized substrates. Different substrates exhibit significant disparities in performance. This paper systematically summarizes the characteristics of different substrates and their ability to remove various contaminants in CWs. For the purposes of practical engineering, this paper provides important reference information for the selection and configuration of substrates.
3. CWs rely on the synergistic interaction of plants, microorganisms, and substrate to purify contaminants. Among them, the substrate plays an important role in CWs. It not only can help microbial proliferation and plant growth, but also directly participate in removing contaminants through filtration and interception, adsorption, chemical precision, redox reactions, and ion exchange.
4. With the continuous improvement in requirements for contaminant removal, EPs have received extensive attention. At present, CWs have been proved to be an effective way to remove EPs. In the future, the mechanism of their removal from CWs should be studied systematically.

Author Contributions: Conceptualization, L.W., L.M. and J.W.; validation, L.W., X.Z. and C.L. (Changqing Liu); formal analysis, X.Z.; investigation, C.L. (Cang Li) and C.J.; resources, Y.J.; data curation, M.X.; writing—original draft preparation, L.W., J.W. and C.L. (Cang Li); writing—review and editing, Y.J. and M.X.; visualization, L.M.; supervision, Y.X.; project administration, C.L. (Cang Li) and Y.X.; funding acquisition, C.L. (Changqing Liu), C.J. and M.X. All authors have read and agreed to the published version of the manuscript.

Funding: This work was supported by Zhangcun River Water Purification Plant Ecological Water Supplement Ecological Health Risk Assessment Technology Research and Development Project (20205541). The Qingdao Science and Technology Demonstration Project for Benefiting the People (No. 23-2-7-zdfn-2-nsh), Taishan Scholar Foundation of Shandong Province (No. tsqn201909126) and the National Key Research and Development Program of China (2021YFC3201004) are also acknowledged.

Data Availability Statement: All data generated or analyzed during this study are included in this published article.

Conflicts of Interest: Authors Liyan Wang, Leihui Ma, Junke Wang, Xia Zhao and Yushu Jing were employed by the company Qingdao Zhangcun River Water Co., Ltd. The remaining authors declare that the research was conducted in the absence of any commercial or financial relationships that could be construed as a potential conflict of interest.

References

1. Sandoval-Herazo, L.C.; Alvarado-Lassman, A.A.; Marín-Muñiz, J.L.; Méndez-Contreras, J.M.; Zamora-Castro, S.A. Effects of the Use of Ornamental Plants and Different Substrates in the Removal of Wastewater Pollutants through Microcosms of Constructed Wetlands. *Sustainability* **2018**, *10*, 1594. [CrossRef]
2. Tian, T.; Yang, Q.; Wei, G.; Cheung, S.G.; Shin, P.K.; Wong, Y.S.; Li, Z.; Chen, Z.; Tam, N.F.Y. Changes of substrate microbial biomass and community composition in a constructed mangrove wetland for municipal wastewater treatment during 10-years operation. *Mar. Pollut. Bull.* **2020**, *155*, 111095. [CrossRef] [PubMed]
3. Lian, J.J.; Xu, S.G.; Zhang, Y.M.; Han, C.W. Molybdenum (VI) removal by using constructed wetlands with different filter media and plants. *Water Sci. Technol.* **2013**, *67*, 1859–1866. [CrossRef] [PubMed]
4. Zeng, Y.; Xu, W.; Wang, H.; Zhao, D.; Ding, H. Nitrogen and Phosphorus Removal Efficiency and Denitrification Kinetics of Different Substrates in Constructed Wetland. *Water* **2022**, *14*, 1757. [CrossRef]
5. Li, X.; Guo, Q.; Wang, Y.; Xu, J.; Wei, Q.; Chen, L.; Liao, L. Enhancing Nitrogen and Phosphorus Removal by Applying Effective Microorganisms to Constructed Wetlands. *Water* **2020**, *12*, 2443. [CrossRef]
6. Negi, D.; Verma, S.; Singh, S.; Daverey, A.; Lin, J.-G. Nitrogen removal via anammox process in constructed wetland—A comprehensive review. *Chem. Eng. J.* **2022**, *437*, 135434. [CrossRef]
7. Shehzadi, M.; Afzal, M.; Khan, M.U.; Islam, E.; Mobin, A.; Anwar, S.; Khan, Q.M. Enhanced degradation of textile effluent in constructed wetland system using *Typha domingensis* and textile effluent-degrading endophytic bacteria. *Water Res.* **2014**, *58*, 152–159. [CrossRef]
8. Türker, O.C.; Böcük, H.; Yakar, A. The phytoremediation ability of a polyculture constructed wetland to treat boron from mine effluent. *J. Hazard. Mater.* **2013**, *252–253*, 132–141. [CrossRef] [PubMed]
9. Song, H.-L.; Nakano, K.; Taniguchi, T.; Nomura, M.; Nishimura, O. Estrogen removal from treated municipal effluent in small-scale constructed wetland with different depth. *Bioresour. Technol.* **2009**, *100*, 2945–2951. [CrossRef]
10. Alsubih, M.; El Morabet, R.; Khan, R.A.; Khan, N.A.; Khan, A.R.; Khan, S.; Mubarak, N.M.; Dehghani, M.H.; Singh, L. Field performance investigation for constructed wetland clubbed with tubesettler for hospital wastewater treatment. *J. Water Process. Eng.* **2022**, *49*, 103147. [CrossRef]
11. Wang, M.; Zhang, D.; Dong, J.; Tan, S.K. Application of constructed wetlands for treating agricultural runoff and agro-industrial wastewater: A review. *Hydrobiologia* **2017**, *805*, 1–31. [CrossRef]
12. Vasilachi, I.C.; Asiminicesei, D.M.; Fertu, D.I.; Gavrilescu, M. Occurrence and Fate of Emerging Pollutants in Water Environment and Options for Their Removal. *Water* **2021**, *13*, 181. [CrossRef]
13. Xiao, J.; Huang, J.; Huang, M.; Chen, M.; Wang, M. Application of basalt fiber in vertical flow constructed wetland for different pollution loads wastewater: Performance, substrate enzyme activity and microorganism community. *Bioresour. Technol.* **2020**, *318*, 124229. [CrossRef] [PubMed]
14. Wei, T.; Zhao, Y.; Guo, J.; Ji, B.; García, A.P.; Núñez, A.E. Developing a novel lightweight substrate for constructed treatment wetland: The idea and the reality. *J. Water Process. Eng.* **2024**, *57*, 104587. [CrossRef]
15. Li, H.Z.; Wang, S.; Ye, J.F.; Xu, Z.X.; Jin, W. A practical method for the restoration of clogged rural vertical subsurface flow constructed wetlands for domestic wastewater treatment using earthworm. *Water Sci. Technol.* **2011**, *63*, 283–290. [CrossRef] [PubMed]
16. Ji, Z.; Tang, W.; Pei, Y. Constructed wetland substrates: A review on development, function mechanisms, and application in contaminants removal. *Chemosphere* **2021**, *286*, 131564. [CrossRef] [PubMed]
17. Wu, H.; Fan, J.; Zhang, J.; Ngo, H.H.; Guo, W.; Liang, S.; Lv, J.; Lu, S.; Wu, W.; Wu, S. Intensified organics and nitrogen removal in the intermittent-aerated constructed wetland using a novel sludge-ceramsite as substrate. *Bioresour. Technol.* **2016**, *210*, 101–107. [CrossRef] [PubMed]

18. Huo, J.Y.; Hu, X.J.; Cheng, S.Y.; Xie, H.J.; Hu, Z.; Wu, H.M.; Liang, S. Effects and mechanisms of constructed wetlands with different substrates on N_2O emission in wastewater treatment. *Environ. Sci. Pollut. Res.* **2022**, *29*, 19045–19053. [CrossRef] [PubMed]
19. Rehrah, D.; Reddy, M.; Novak, J.; Bansode, R.; Schimmel, K.; Yu, J.; Watts, D.; Ahmedna, M. Production and characterization of biochars from agricultural by-products for use in soil quality enhancement. *J. Anal. Appl. Pyrolysis* **2014**, *108*, 301–309. [CrossRef]
20. Saeed, T.; Muntaha, S.; Rashid, M.; Sun, G.; Hasnat, A. Industrial wastewater treatment in constructed wetlands packed with construction materials and agricultural by-products. *J. Clean. Prod.* **2018**, *189*, 442–453. [CrossRef]
21. Nandhini, K.; Karthikeyan, J. Influence of Industrial and Agricultural by-Products as Cementitious Blends in Self-Compacting Concrete—A Review. *Silicon* **2021**, *14*, 2431–2452. [CrossRef]
22. Ke, Y.; Chen, Y.; Liang, S.; Hu, J.; Hou, H.; Quan, J.; Li, X.; Duan, H.; Yuan, S.; Yang, J. Environmentally sound management of industrial solid waste: A paradigm of proposed bi-tetrahedron. *Resour. Conserv. Recycl.* **2023**, *198*, 107212. [CrossRef]
23. Samanlioglu, F. A multi-objective mathematical model for the industrial hazardous waste location-routing problem. *Eur. J. Oper. Res.* **2013**, *226*, 332–340. [CrossRef]
24. Sahu, A.; Kumar, S.; Srivastava, A.; Pratap, B. Performance of recycled aggregate concrete using copper slag as fine aggregate. *J. Build. Eng.* **2024**, *82*, 108364. [CrossRef]
25. Liu, W.; Chen, X.; Li, W.; Yu, Y.; Yan, K. Environmental assessment, management and utilization of red mud in China. *J. Clean. Prod.* **2014**, *84*, 606–610. [CrossRef]
26. Yang, C.; Zhang, X.; Tang, Y.; Jiang, Y.; Xie, S.; Zhang, Y.; Qin, Y. Selection and optimization of the substrate in constructed wetland: A review. *J. Water Process. Eng.* **2022**, *49*, 103140. [CrossRef]
27. Kong, F.; Zhang, Y.; Wang, H.; Tang, J.; Li, Y.; Wang, S. Removal of Cr(VI) from wastewater by artificial zeolite spheres loaded with nano Fe–Al bimetallic oxide in constructed wetland. *Chemosphere* **2020**, *257*, 127224. [CrossRef] [PubMed]
28. Ge, Y.; Wang, X.; Zheng, Y.; Dzakpasu, M.; Zhao, Y.; Xiong, J. Functions of slags and gravels as substrates in large-scale demonstration constructed wetland systems for polluted river water treatment. *Environ. Sci. Pollut. Res.* **2015**, *22*, 12982–12991. [CrossRef] [PubMed]
29. Zhong, H.; Hu, N.; Wang, Q.H.; Chen, Y.C.; Huang, L. How to select substrate for alleviating clogging in the subsurface flow constructed wetland? *Sci. Total Environ.* **2022**, *828*, 154529. [CrossRef] [PubMed]
30. Wang, Z.; Dong, J.; Liu, L.; Zhu, G.; Liu, C. Screening of phosphate-removing substrates for use in constructed wetlands treating swine wastewater. *Ecol. Eng.* **2013**, *54*, 57–65. [CrossRef]
31. Cuomo, M.; König, R.; Zanardini, E.; Di Guardo, A.; Bianchi, G.; Ortona, A.; Principi, P. Using zeolite filters to reduce activated carbon use in micropollutant removal from wastewater. *J. Water Process. Eng.* **2023**, *56*, 104298. [CrossRef]
32. Velarde, L.; Nabavi, M.S.; Escalera, E.; Antti, M.-L.; Akhtar, F. Adsorption of heavy metals on natural zeolites: A review. *Chemosphere* **2023**, *328*, 138508. [CrossRef] [PubMed]
33. Jiang, P.; Zhou, L.; Wang, W.; Li, N.; Zhang, F. Performance and mechanisms of fly ash for graphene oxide removal from aqueous solution. *Environ. Sci. Pollut. Res.* **2021**, *29*, 3773–3783. [CrossRef] [PubMed]
34. Zhang, J.; Zou, Y.; Yu, X.; Ding, S.; Yan, J.; Min, Y. Vegetated Steel Slag Substrate Constructed Wetlands can Achieve High Efficiency Simultaneous Nitrogen and Phosphorus Removal. *Front. Environ. Sci.* **2022**, *10*, 947783. [CrossRef]
35. Yang, Y.; Wang, Z.M.; Liu, C.; Guo, X.C. Enhanced P, N and C removal from domestic wastewater using constructed wetland employing construction solid waste (CSW) as main substrate. *Water Sci. Technol.* **2012**, *66*, 1022–1028. [CrossRef] [PubMed]
36. Lu, J.; Dong, L.; Guo, Z.; Hu, Z.; Dai, P.; Zhang, J.; Wu, H. Highly efficient phosphorous removal in constructed wetland with iron scrap: Insights into the microbial removal mechanism. *J. Environ. Manag.* **2023**, *347*, 119076. [CrossRef] [PubMed]
37. Deng, S.; Li, D.; Yang, X.; Xing, W.; Li, J.; Zhang, Q. Iron [Fe(0)]-rich substrate based on iron–carbon micro–electrolysis for phosphorus adsorption in aqueous solutions. *Chemosphere* **2017**, *168*, 1486–1493. [CrossRef] [PubMed]
38. Carstensen, M.V.; Larsen, S.E.; Kjærgaard, C.; Hoffmann, C.C. Reducing adverse side effects by seasonally lowering nitrate removal in subsurface flow constructed wetlands. *J. Environ. Manag.* **2019**, *240*, 190–197. [CrossRef] [PubMed]
39. Chaves-Barquero, L.G.; Humeniuk, B.W.; Luong, K.H.; Cicek, N.; Wong, C.S.; Hanson, M.L. Crushed recycled glass as a substrate for constructed wetland wastewater treatment: A case study of its potential to facilitate pharmaceutical removal. *Environ. Sci. Pollut. Res.* **2021**, *28*, 52306–52318. [CrossRef] [PubMed]
40. Wan, Q.; Han, Q.; Luo, H.; He, T.; Xue, F.; Ye, Z.; Chen, C.; Huang, S. Ceramsite Facilitated Microbial Degradation of Pollutants in Domestic Wastewater. *Int. J. Environ. Res. Public Health* **2020**, *17*, 4692. [CrossRef] [PubMed]
41. Vispo, C.; Geronimo, F.K.; Jeon, M.; Kim, L.-H. Performance Evaluation of Various Filter Media for Multi-Functional Purposes to Urban Constructed Wetlands. *Sustainability* **2023**, *16*, 287. [CrossRef]
42. Gupta, P.; Ann, T.-W.; Lee, S.-M. Use of biochar to enhance constructed wetland performance in wastewater reclamation. *Environ. Eng. Res.* **2016**, *21*, 36–44. [CrossRef]
43. Yun, Y.; Zhou, X.; Li, Z.; Uddin, S.M.N.; Bai, X. Comparative research on phosphorus removal by pilot-scale vertical flow constructed wetlands using steel slag and modified steel slag as substrates. *Water Sci. Technol.* **2015**, *71*, 996–1003. [CrossRef] [PubMed]
44. Zhu, H.; Yan, B.; Xu, Y.; Guan, J.; Liu, S. Removal of nitrogen and COD in horizontal subsurface flow constructed wetlands under different influent C/N ratios. *Ecol. Eng.* **2014**, *63*, 58–63. [CrossRef]

45. Xu, Q.; Cui, L. Removal of COD from synthetic wastewater in vertical flow constructed wetland. *Water Environ. Res.* **2019**, *91*, 1661–1668. [CrossRef] [PubMed]
46. Wang, Y.; Yang, J.; Xu, H.; Liu, C.; Shen, Z.; Hu, K. Preparation of Ceramsite Based on Waterworks Sludge and Its Application as Matrix in Constructed Wetlands. *Int. J. Environ. Res. Public Health* **2019**, *16*, 2637. [CrossRef] [PubMed]
47. Wang, H.X.; Wang, X.Y.; Teng, H.W.; Xu, J.L.; Sheng, L.X. Purification mechanism of city tail water by constructed wetland substrate with NaOH-modified corn straw biochar. *Ecotoxicol. Environ. Saf.* **2022**, *238*, 113597. [CrossRef] [PubMed]
48. Wang, H.; Xu, J.; Sheng, L.; Teng, H. Study on treatment of city tail water by constructed wetland with corn straw biochar substrate. *Environ. Technol. Innov.* **2022**, *28*, 102855. [CrossRef]
49. Kong, F.L.; Wang, J.R.; Hou, W.H.; Cui, Y.Q.; Yu, L.H.; Zhang, Y.; Wang, S. Influence of modified biochar supported sulfidation of nano-zero-valent-iron (S-nZVI/BC) on nitrate removal and greenhouse gas emission in constructed wetland. *J. Environ. Sci.* **2023**, *125*, 568–581. [CrossRef] [PubMed]
50. Liu, Y.; Liu, X.-H.; Wang, H.-C.; Li, Z.-L.; Liang, B.; Sun, Y.-L.; Cheng, H.-Y.; Lu, S.-Y.; Wang, A.-J. Pyrite coupled with steel slag to enhance simultaneous nitrogen and phosphorus removal in constructed wetlands. *Chem. Eng. J.* **2023**, *470*, 143944. [CrossRef]
51. Zheng, T.; Lin, X.; Xu, J.; Ren, J.; Sun, D.; Gu, Y.; Huang, J. Enhanced Nitrogen Removal of Steel Rolling Wastewater by Constructed Wetland Combined with Sulfur Autotrophic Denitrification. *Sustainability* **2021**, *13*, 1559. [CrossRef]
52. Chu, Y.; Liu, W.; Tan, Q.; Yang, L.; Chen, J.; Ma, L.; Zhang, Y.; Wu, Z.; He, F. Vertical-flow constructed wetland based on pyrite intensification: Mixotrophic denitrification performance and mechanism. *Bioresour. Technol.* **2022**, *347*, 126710. [CrossRef]
53. Wang, W.; Song, X.; Li, F.; Ji, X.; Hou, M. Intensified nitrogen removal in constructed wetlands by novel spray aeration system and different influent COD/N ratios. *Bioresour. Technol.* **2020**, *306*, 123008. [CrossRef] [PubMed]
54. Zhou, X.; Liang, C.; Jia, L.; Feng, L.; Wang, R.; Wu, H. An innovative biochar-amended substrate vertical flow constructed wetland for low C/N wastewater treatment: Impact of influent strengths. *Bioresour. Technol.* **2018**, *247*, 844–850. [CrossRef] [PubMed]
55. Inglezakis, V.J. The concept of "capacity" in zeolite ion-exchange systems. *J. Colloid Interface Sci.* **2005**, *281*, 68–79. [CrossRef]
56. Alshameri, A.; Ibrahim, A.; Assabri, A.M.; Lei, X.; Wang, H.; Yan, C. The investigation into the ammonium removal performance of Yemeni natural zeolite: Modification, ion exchange mechanism, and thermodynamics. *Powder Technol.* **2014**, *258*, 20–31. [CrossRef]
57. Tang, X.Q.; Huang, S.L.; Scholz, M. Comparison of phosphorus removal between vertical subsurface flow constructed wetlands with different substrates. *Water Environ. J.* **2009**, *23*, 180–188. [CrossRef]
58. Jakubaszek, A. Nitrogen and Phosphorus Accumulation in Horizontal Subsurface Flow Constructed Wetland. *Agronomy* **2021**, *11*, 1317. [CrossRef]
59. Le, T.; Ngo, H.; Guo, W.; Vu, N.; Tran, T.; Nguyen, T.; Nguyen, X.; Nguyen, V.; Pham, T. Hybrid use of coal slag and calcined ferralsol as wetland substrate for improving phosphorus removal from wastewater. *Chem. Eng. J.* **2022**, *428*, 132124. [CrossRef]
60. Han, C.; Wang, Z.; Yang, W.; Wu, Q.; Yang, H.; Xue, X. Effects of pH on phosphorus removal capacities of basic oxygen furnace slag. *Ecol. Eng.* **2016**, *89*, 1–6. [CrossRef]
61. Jiang, C.; Jia, L.; Zhang, B.; He, Y.; Kirumba, G. Comparison of quartz sand, anthracite, shale and biological ceramsite for adsorptive removal of phosphorus from aqueous solution. *J. Environ. Sci.* **2014**, *26*, 466–477. [CrossRef] [PubMed]
62. Guaya, D.; Valderrama, C.; Farran, A.; Armijos, C.; Cortina, J.L. Simultaneous phosphate and ammonium removal from aqueous solution by a hydrated aluminum oxide modified natural zeolite. *Chem. Eng. J.* **2015**, *271*, 204–213. [CrossRef]
63. Stocker, K.; Ellersdorfer, M. Phosphate Fixation and P Mineralogy on Natural and Ca-Modified Zeolites During Simultaneous Nutrient Removal. *Water Air Soil Pollut.* **2022**, *233*, 41. [CrossRef]
64. Zaimee, M.Z.A.; Sarjadi, M.S.; Rahman, L. Heavy Metals Removal from Water by Efficient Adsorbents. *Water* **2021**, *13*, 2659. [CrossRef]
65. Le, T.V.; Nguyen, B.T. Heavy metal pollution in surface water bodies in provincial Khanh Hoa, Vietnam: Pollution and human health risk assessment, source quantification, and implications for sustainable management and development. *Environ. Pollut.* **2024**, *343*, 123216. [CrossRef] [PubMed]
66. Bavandpour, F.; Zou, Y.; He, Y.; Saeed, T.; Sun, Y.; Sun, G. Removal of dissolved metals in wetland columns filled with shell grits and plant biomass. *Chem. Eng. J.* **2018**, *331*, 234–241. [CrossRef]
67. Hafeznezami, S.; Kim, J.-L.; Redman, J. Evaluating Removal Efficiency of Heavy Metals in Constructed Wetlands. *J. Environ. Eng.* **2012**, *138*, 475–482. [CrossRef]
68. Šíma, J.; Svoboda, L.; Pomijová, Z. Removal of Selected Metals from Wastewater Using a Constructed Wetland. *Chem. Biodivers.* **2016**, *13*, 582–590. [CrossRef] [PubMed]
69. Dehmani, Y.; Mohammed, B.B.; Oukhrib, R.; Dehbi, A.; Lamhasni, T.; Brahmi, Y.; El-Kordy, A.; Franco, D.S.; Georgin, J.; Lima, E.C.; et al. Adsorption of various inorganic and organic pollutants by natural and synthetic zeolites: A critical review. *Arab. J. Chem.* **2024**, *17*, 105474. [CrossRef]
70. Lee, H.-S.; Shin, H.-S. Competitive adsorption of heavy metals onto modified biochars: Comparison of biochar properties and modification methods. *J. Environ. Manag.* **2021**, *299*, 113651. [CrossRef] [PubMed]
71. Tang, S.-F.; Zhou, H.; Tan, W.-T.; Huang, J.-G.; Zeng, P.; Gu, J.-F.; Liao, B.-H. Adsorption Characteristics and Mechanisms of Fe-Mn Oxide Modified Biochar for Pb(II) in Wastewater. *Int. J. Environ. Res. Public Health* **2022**, *19*, 8420. [CrossRef] [PubMed]
72. Otal, E.H.; Kim, M.L.; Dietrich, S.; Takada, R.; Nakaya, S.; Kimura, M. Open-Source Portable Device for the Determination of Fluoride in Drinking Water. *ACS Sensors* **2021**, *6*, 259–266. [CrossRef] [PubMed]

73. Styburski, D.; Baranowska-Bosiacka, I.; Goschorska, M.; Chlubek, D.; Gutowska, I. Beer as a Rich Source of Fluoride Delivered into the Body. *Biol. Trace Element Res.* **2016**, *177*, 404–408. [CrossRef] [PubMed]
74. Yao, D.D.; Hu, X.J.; Shen, X.T.; Xie, H.J.; Hu, Z.; Zhang, J.; Liang, S. Can calcium-based constructed wetlands improve fluoride removal performance? *Chem. Eng. J.* **2022**, *450*, 138314. [CrossRef]
75. Rickard, B.P.; Overchuk, M.; Tulino, J.; Tan, X.; Ligler, F.S.; Bae-Jump, V.L.; Fenton, S.E.; Rizvi, I. Exposure to select PFAS and PFAS mixtures alters response to platinum-based chemotherapy in endometrial cancer cell lines. *Environ. Health* **2023**, *22*, 87. [CrossRef] [PubMed]
76. Ma, H.; Kang, Y.; Li, M.; Dong, J.; Wang, Y.; Xiao, J.; Guo, Z. Enhancement of perfluorooctanoic acid and perfluorooctane sulphonic acid removal in constructed wetland using iron mineral: Performance and mechanisms. *J. Hazard. Mater.* **2023**, *447*, 130819. [CrossRef] [PubMed]
77. Srain, H.S.; Beazley, K.F.; Walker, T.R. Pharmaceuticals and personal care products and their sublethal and lethal effects in aquatic organisms. *Environ. Rev.* **2021**, *29*, 142–181. [CrossRef]
78. Gilroy, E.A.M.; Klinck, J.S.; Campbell, S.D.; McInnis, R.; Gillis, P.L.; de Solla, S.R. Toxicity and bioconcentration of the pharmaceuticals moxifloxacin, rosuvastatin, and drospirenone to the unionid mussel *Lampsilis siliquoidea*. *Sci. Total Environ.* **2014**, *487*, 537–544. [CrossRef] [PubMed]
79. Krogh, J.; Lyons, S.; Lowe, C.J. Pharmaceuticals and Personal Care Products in Municipal Wastewater and the Marine Receiving Environment Near Victoria Canada. *Front. Mar. Sci.* **2017**, *4*, 415. [CrossRef]
80. Chen, J.; Deng, W.-J.; Liu, Y.-S.; Hu, L.-X.; He, L.-Y.; Zhao, J.-L.; Wang, T.-T.; Ying, G.-G. Fate and removal of antibiotics and antibiotic resistance genes in hybrid constructed wetlands. *Environ. Pollut.* **2019**, *249*, 894–903. [CrossRef] [PubMed]
81. Liu, Y.; Liu, X.; Lu, S.; Zhao, B.; Wang, Z.; Xi, B.; Guo, W. Adsorption and biodegradation of sulfamethoxazole and ofloxacin on zeolite: Influence of particle diameter and redox potential. *Chem. Eng. J.* **2019**, *384*, 123346. [CrossRef]
82. Rodriguez, E.; Campinas, M.; Acero, J.L.; Rosa, M.J. Investigating PPCP Removal from Wastewater by Powdered Activated Carbon/Ultrafiltration. *Water Air Soil Pollut.* **2016**, *227*, 177. [CrossRef]
83. Li, J.; Zhang, K.; Zhang, H. Adsorption of antibiotics on microplastics. *Environ. Pollut.* **2018**, *237*, 460–467. [CrossRef] [PubMed]
84. Wang, W.; Ge, J.; Yu, X. Bioavailability and toxicity of microplastics to fish species: A review. *Ecotoxicol. Environ. Saf.* **2020**, *189*, 109913. [CrossRef] [PubMed]
85. Ahmed, R.; Hamid, A.K.; Krebsbach, S.A.; He, J.; Wang, D. Critical review of microplastics removal from the environment. *Chemosphere* **2022**, *293*, 133557. [CrossRef] [PubMed]
86. Zhao, Y.F.; Jin, R.X.; Chen, Y.H.; Zhang, J.H.; Tao, S.Y.; Liu, S.W.; Shen, M.C. Constructed wetlands as neglected fixed source of microplastics and antibiotic resistance genes in natural water bodies? *Sci. Total Environ.* **2023**, *902*, 166474. [CrossRef] [PubMed]
87. Chen, Y.; Li, T.; Hu, H.; Ao, H.; Xiong, X.; Shi, H.; Wu, C. Transport and fate of microplastics in constructed wetlands: A microcosm study. *J. Hazard. Mater.* **2021**, *415*, 125615. [CrossRef] [PubMed]
88. Shen, M.; Hu, T.; Huang, W.; Song, B.; Zeng, G.; Zhang, Y. Removal of microplastics from wastewater with aluminosilicate filter media and their surfactant-modified products: Performance, mechanism and utilization. *Chem. Eng. J.* **2021**, *421*, 129918. [CrossRef]
89. Miranda, S.T.; de Matos, A.T.; de Matos, M.P.; Saraiva, C.B.; Teixeira, D.L. Influence of the substrate type and position of plant species on clogging and the hydrodynamics of constructed wetland systems. *J. Water Process. Eng.* **2019**, *31*, 100871. [CrossRef]
90. Wang, H.; Sun, J.; Xu, J.; Sheng, L. Study on clogging mechanisms of constructed wetlands from the perspective of wastewater electrical conductivity change under different substrate conditions. *J. Environ. Manag.* **2021**, *292*, 112813. [CrossRef]
91. Zhou, Y.; Gu, T.; Yi, W.; Zhang, T.; Zhang, Y. The release mechanism of heavy metals from lab-scale vertical flow constructed wetlands treating road runoff. *Environ. Sci. Pollut. Res.* **2019**, *26*, 16588–16595. [CrossRef] [PubMed]
92. Ury, E.A.; Arrumugam, P.; Herbert, E.R.; Badiou, P.; Page, B.; Basu, N.B. Source or sink? Meta-analysis reveals diverging controls of phosphorus retention and release in restored and constructed wetlands. *Environ. Res. Lett.* **2023**, *18*, 083002. [CrossRef]
93. Ávila, C.; Bayona, J.M.; Martín, I.; Salas, J.J.; García, J. Emerging organic contaminant removal in a full-scale hybrid constructed wetland system for wastewater treatment and reuse. *Ecol. Eng.* **2015**, *80*, 108–116. [CrossRef]
94. Jehawi, O.H.; Abdullah, S.R.S.; Kurniawan, S.B.; Ismail, N.; Idris, M.; Al Sbani, N.H.; Muhamad, M.H.; Abu Hasan, H. Performance of pilot Hybrid Reed Bed constructed wetland with aeration system on nutrient removal for domestic wastewater treatment. *Environ. Technol. Innov.* **2020**, *19*, 100891. [CrossRef]
95. Sharma, P.K.; Rausa, K.; Rani, A.; Mukherjee, S.; Kumar, M. Biopurification of dairy farm wastewater through hybrid constructed wetland system: Groundwater quality and health implications. *Environ. Res.* **2021**, *200*, 111426. [CrossRef]
96. Jiang, S.; Xu, J.; Wang, H.; Wang, X. Study of the effect of pyrite and alkali-modified rice husk substrates on enhancing nitrogen and phosphorus removals in constructed wetlands. *Environ. Sci. Pollut. Res.* **2022**, *29*, 54234–54249. [CrossRef] [PubMed]
97. Gupta, S.; Srivastava, P.; Patil, S.A.; Yadav, A.K. A comprehensive review on emerging constructed wetland coupled microbial fuel cell technology: Potential applications and challenges. *Bioresour. Technol.* **2021**, *320*, 124376. [CrossRef] [PubMed]
98. Chen, K.-C.; Wang, Y.-H.; Lu, Y.-C. Treatment of polluted water for reclamation using photocatalysis and constructed wetlands. *Catal. Today* **2011**, *175*, 276–282. [CrossRef]

99. Zhang, Y.; Liu, F.; Lin, Y.; Sun, L.; Guo, X.; Yang, S.; He, J. Enhanced Swine Wastewater Treatment by Constructed Wetland—Microbial Fuel Cell Systems. *Water* **2022**, *14*, 3930. [CrossRef]
100. Xiao, E.-R.; Liang, W.; He, F.; Cheng, S.-P.; Wu, Z.-B. Performance of the combined SMBR–IVCW system for wastewater treatment. *Desalination* **2010**, *250*, 781–786. [CrossRef]

water

MDPI

Review

Coupling between Increased Amounts of Microplastics and Dissolved Organic Compounds in Water

Wengang Yan [1], Qianjin Wang [1], Ya Gao [2], Mengchen Xu [1], Huiying Li [1], Yuping Zhou [1], Changqing Liu [1] and Yihua Xiao [1,*]

[1] School of Environmental and Municipal Engineering, Qingdao University of Technology, Qingdao 266520, China
[2] College of Municipal and Ecological Engineering, Shanghai Urban Construction Vocational College, Shanghai 200438, China
* Correspondence: yihua.xiao@qut.edu.cn

Abstract: Microplastic (MP) pollution is a rapidly spreading global problem, threatening the use and sustainability of freshwater resources. MPs in water can act as both a source and sink of dissolved organic compounds. This review summarizes the current knowledge of interactions between MPs and dissolved organic compounds, including the adsorption and release of dissolved organic compounds by MPs and the impacts of MPs on the source and sink of natural dissolved organic matter (DOM) in aquatic ecosystems. The key mechanisms for the adsorption of dissolved organic compounds on MPs are hydrophobic interactions, van der Waals forces, and π–π interactions. Particle size, morphological characteristics, density, and environmental factors (pH, ionic strength, and UV radiation) have a great influence on the adsorption of dissolved organic compounds on MPs. Although research on the interactions between dissolved organic compounds and MPs has progressed rapidly, to date, research on the impacts of increasing amounts of MPs on natural DOM cycles (production, transformation, and fate) in aquatic ecosystems has been very limited. Knowledge gaps and future research directions are outlined at the end of this review.

Keywords: microplastics; adsorption; release; dissolved organic matter; source; sink

1. Introduction

With the growth of the plastic industry all over the world, plastic pollution in the environment has drawn much attention [1–3]. Currently, plastic debris is widely found in the ocean [4,5], freshwater [1,6], sewage [7], sediments [8], and aquatic organisms [9]. Increasing amounts of plastic in water have pronounced effects on freshwater ecology and societal services, as increasing plastic pollution may have toxic effects on aquatic organisms [10], alter aquatic species composition [11], and increase the cost of drinking water purification [12]. Microplastics (MPs) are defined as plastic debris with a diameter of less than 5 mm [13], characterized by a small size and large specific surface area. Based on their chemical composition, MPs are classified as polyethylene (PE), polypropylene (PP), polyamide (PA), polyvinyl chloride (PVC), polystyrene (PS), polyethylene terephthalate (PET), etc. There are two primary sources of MPs in aquatic systems: (1) fragmentation of plastic litter released into water through solar radiation and microbial degradation [14,15], and (2) direct discharge from municipal wastewater treatment plants, which may contain personal care and cosmetic products or textile fiber residues [7].

MPs in the environment are often hydrophobic, chemically stable, and resistant to degradation [16]. However, studies have shown that MPs can leach polymers or additives into the surrounding environment through biotic and abiotic processes [17–19]. Plastic additives are chemical compounds incorporated into plastic during the manufacturing process, which can be used as plasticizers, foaming agents, flame retardants, antioxidants, stabilizers, and pigments [20]. Most plastic additives are hydrophobic organic compounds and are recognized as toxic to aquatic organisms [21]. Lithner et al. (2009) tested the acute toxicity of plastic leachates to *Daphnia magna* and found that 9 of 32 plastic leachates had different toxic effects (48 h EC_{50}s ranging from 5 to 80 g plastic L^{-1}) on *Daphnia* mobility [22]. After losing the protection of additives, the degradation of MPs in water can be accelerated [23]. There is a great possibility that MP leachates (additives or any other organic substrate) released into aquatic systems contribute to carbon cycling in water. However, the effects of MP pollution on carbon cycling in water are not well studied.

Once released into the aquatic ecosystem, MPs can serve as a carrier for various organic pollutants and heavy metals [13,24], affecting the transportation and fate of chemical pollutants. Ziajahromi et al. (2019) studied the effect of PE MPs on the acute toxicity of the synthetic pyrethroid bifenthrin to the invertebrate *Chironomus tepperi* in both synthetic water and river water and found that the presence of PE MPs could reduce the toxicity of bifenthrin to exposed larvae [25]. Studies have shown that PVC, PE, PP, and PS have a high sorption capacity for organic pollutants, such as dichlorodiphenyltrichloroethane (DDT), phenanthrene (Phe), bis-2-Ethylhexyl phthalate (DEHP), and chlorinated benzenes [26–28]. The adsorption by MPs can transfer chemical compounds from water into aquatic organisms. The concentrations of organic pollutants (polychlorinated biphenyls and polybrominated diphenyl ethers) in aquatic organisms have been found to be positively related to the amount of ingested MPs [29,30]. However, the adsorption capacity for organic pollutants on MPs highly depends on the MPs' characteristics, the chemical composition of the organic pollutants, and environmental factors, such as pH and ionic strength [28].

Besides adsorption, MPs can leach organic compounds into water. Most of the plastic additives are not covalently bound to MPs and can be released into the surrounding water. The release process can be assisted by solar radiation and the biological degradation of MPs [31]. DEHP and bisphenol A (BPA) additives have been detected in commercial PVC and PS plastic solutions under UV light [17,32]. Polybrominated diphenyl ethers (PBDEs), phthalates, nonylphenols (NPs), and antioxidant additives have also been found in marine environments [20]. Besides additives, MPs can leach other dissolved organic compounds derived from MP polymers into the aqueous phase. A few published studies have documented that these polymer-derived dissolved organic compounds potentially contribute to dissolved organic matter (DOM) pools in aquatic ecosystems [23,33]. DOM is a complex mixture of organic molecules that can pass through a 0.2–0.7 μm filter [34]. MPs in natural waters can form a "micro-environment", the chemical composition and microbial activities of which are significantly different from those of the surrounding environment. This "micro-environment" can affect the biogeochemical cycles of natural DOM and microorganic pollutants [23,35–37]. MPs can also serve as a carrier for DOM, which significantly impacts DOM transport along the aquatic continuum [38].

Research on the occurrence and ecological effects of MPs in water has progressed rapidly in recent years. This study aims to review the current research and identify knowledge gaps in the interactions between increasing amounts of MPs and dissolved organic compounds in aquatic systems. In Sections 2 and 3, we summarize the adsorption and release of dissolved organic compounds by MPs and their environmental factors. In Section 4, we explore the current knowledge on the impacts of MPs on the source and sink of natural DOM. Guidance on future study directions in this field is provided at the end of this paper.

2. Adsorption of Dissolved Organic Compounds on MPs

Currently, studies on the adsorption of dissolved organic compounds by MPs mainly focus on the following aspects: (1) the adsorption mechanisms of dissolved organic compounds on MPs, and (2) the impact of environmental factors on the adsorption of dissolved organic compounds on MPs. Studies on the adsorption of dissolved organic compounds on MPs are summarized in Table 1.

Table 1. The adsorption mechanisms of dissolved organic compounds on MPs and the influencing factors.

Adsorption Mechanisms	Influencing Factors	Specific Process	Literature Source
Hydrophobic interactions	Chemical property	Adsorption increases with increasing hydrophobicity	[27,39–44]
	pH	Increasing pH → promoting dissociation of hydrophobic neutral sorbent molecules into hydrophilic, negatively charged substances → reducing hydrophobic interactions	[42,45]
	Ionic strength	The presence of salt ions → reducing the solubility of the compound in aqueous solution → promoting hydrophobic adsorption of the compound on MPs	[40,42,46,47]
	UV radiation	UV irradiation → leading to the aging process → changing the surface properties (size, surface area, porosity, surface polarity) of MPs → enhancing the adsorption of hydrophilic organic pollutants	[48,49]
Van der Waals force	Temperature	Increasing temperature → increasing the mobility and solubility of the adsorbed molecules → the van der Waals forces decrease → the adsorption capacity decreases	[50]
	Ionic strength	Increasing ion concentrations → ions (Na and Ca) can occupy the adsorption sites of MPs → decreasing absorption capacity	[48]
π–π interactions	pH	Increasing pH → increasing the π donor capacity of the sorbent → enhancing π–π interactions	[51,52]
	Ionic strength	The stronger the cation → the stronger the binding between organic compounds and other ionic compounds → leading to increased adsorption of organic compounds on MPs	[53]
Electrostatic interactions	pH	Negatively charged MP surface and lower pH → leading to the protonation of MPs → contributing to the electrostatic gravitational force between organic compounds and MPs	[43,48,53–55]
	Ionic strength	Electrolytes can compete with adsorbates for electrostatic sites → decreasing the adsorption capacity	[43,53,56,57]

2.1. Adsorption of Dissolved Organic Compounds on MPs: Mechanisms

2.1.1. Hydrophobic Interactions

Hydrophobic interactions are of major importance for the adsorption of nonpolar dissolved organic compounds on MPs in aqueous solutions [38]. Hüffer and Hofmann (2016) investigated the sorption behavior of seven aliphatic and aromatic organic compounds on four types of MPs (PA, PE, PVC, and PS). The experimental isotherm results showed that the chemical properties of both the MPs (monomeric composition) and organic compounds (hydrophobicity) had important effects on the sorption behavior [58]. The adsorption of Suwannee River humic acid (SRHA) and Suwannee River fulvic acid (SRFA) on PS in an aquatic environment was investigated using kinetic, isotherm, and site energy distribution analyses [39]. The results showed that hydrophobic interactions were one of the two main adsorption mechanisms of SRHA and SRFA adsorption on PS. The adsorption capacity for different organic compounds on MPs is positively correlated with their octanol–water partition coefficient (K_{ow} or log K_{ow}), a parameter that is commonly used as a hydrophobic parameter [37,42,59].

2.1.2. Van der Waals Force

For nonpolar aliphatic polymers with no specific functional groups, their interactions with organic compounds are mainly van der Waals interactions. For instance, Guo and Wang (2019) demonstrated that the sorption capacity for sulfamethoxazole (SMX) on polystyrene (PS) was higher than that of polypropylene (PP) [41]. This phenomenon is due to the different chemical characteristics of PE and PS. PE and PS are aliphatic and aromatic polymers, respectively. The adsorption of SMX on PE and PS is driven by van der Waals and π–π interactions, respectively, based on their aliphatic and aromatic characteristics. The higher energy of π–π interactions than that of van der Waals interactions resulted in a higher adsorption capacity for SMX on PS than on PE [41,43]. Li et al. (2018) also showed that the adsorption affinity of sulfadiazine (SDZ) on PS through both non-specific van der Waals and π–π interactions was higher than that on PE through only van der Waals interactions [54].

2.1.3. π–π Interactions

In the summary of the adsorption mechanisms of organic compounds on MPs (Table 1), π–π interactions are often discussed together with van der Waals interactions. Through spectroscopic analyses, Chen et al. (2018) investigated the interaction between commercial dissolved organic matter humic acid (HA) and PS. Their results showed that PS mainly interacts with the aromatic structure of dissolved organic compounds via π–π interactions and is then entrapped in the organic molecule by carboxyl-group C=O bonds within the DOM [52]. The study by Zhang et al. (2012) showed that, in addition to hydrophobic interactions, π–π electron donor–acceptor interactions play an important role in the adsorption of HA and FA on porous PS [53].

2.2. Adsorption of Dissolved Organic Compounds on MPs: The Effect of MP Characteristics

The adsorption of dissolved organic compounds on MPs is highly affected by the characteristics of MPs, such as their particle size, porosity, and specific surface area [60]. The chemical composition of a polymer is described based on its chain structure. The polymer chains are arranged and stacked to form an aggregate structure, including crystalline and amorphous components. The influence of the aggregate structure on adsorption is usually more important than that of chain structures. In addition, other structural properties of MPs, such as their density, degree of crystallinity, and abundance of rubbery and glassy states in the amorphous region, can directly lead to different adsorption capacities [54,61].

2.2.1. Particle Size

The particle size of MPs is of great importance for the adsorption of dissolved organic compounds on MPs [62,63]. Chen et al. (2018) demonstrated that the affinity of MP and DOM interactions is highly dependent on the particle size of the MPs [52]. Smaller MP particle sizes possess a larger specific surface area and higher porosity, which are favorable for the adsorption of DOM on MPs [42,64,65]. Studies show that MPs' specific surface area and adsorption sites increase with decreasing particle size, which can enhance the adsorption capacity for dissolved organic compounds on MPs [40]. However, Fang et al. (2019) found that the adsorption capacity for triazole fungicide was the highest at a PS particle size of 10 μm, followed by a particle size of 2 μm. The reason for this phenomenon may be due to the agglomeration of MPs with particle sizes that are too small [66].

2.2.2. Morphological Effects

Based on molecular chain arrangements, plastic polymers can be classified into crystalline, semi-crystalline, and amorphous structures. Amorphous structures consist of rubber and glass subcomponents. Plastics in a glassy state can transform into a rubbery state at a certain temperature, which can be influenced by a few factors including chain segment fluidity [56]. The morphological transformation of plastics can influence their adsorption of organic compounds [58]. Zhao et al. (2020) found that the adsorption capacity for organic compounds on polar MPs was almost two orders of magnitude higher than that on nonpolar MPs, due to the higher proportion of rubbery states in the amorphous region of the polar MPs [67]. Liu et al. (2019) demonstrated that rubbery PS and PE possess a larger internal cavity volume in the rubbery region compared to their glassy counterparts, which favors the accumulation and adsorption of polycyclic aromatic hydrocarbons (PAHs) [68].

On the other hand, the surface characteristics and crystallinity degree of MPs can determine the adsorption of antibiotics on MPs [54]. Guo et al. (2012) also found that PEs with the same chemical composition but different crystallinities (ranging from 58.7%

to 25.5%) had varying organic carbon content-normalized sorption coefficients (Koc) for hydrophobic organic compounds. However, the study by Liu et al. (2019) found no significant relationship between bisphenol A adsorption capacity and MPs' crystallinity. These studies demonstrated that the sorption capacity between organic compounds and MPs is not only affected by the crystallinity factor but also by other factors including surface functional groups and surface roughness, which can influence the hydrophobicity of MPs and regulate the interactions between organic compounds and MPs [46,69].

2.2.3. Density

The density of MPs also affects the adsorption of dissolved organic compounds on MPs. For instance, low-density PE showed a higher diffusion rate for PAHs than high-density PE, demonstrating that the adsorption of organic compounds on MPs is inversely related to the density of MPs [70]. When studying the sorption behavior of three non-steroidal anti-inflammatory drugs (NSAIDs) on PE MPs with different densities, Elizalde-Velázquez et al. (2020) found that, in comparison with ultrahigh-density PE and medium-density PE, low-density PE had the highest sorption capacity for NSAIDs [71].

2.3. *Adsorption of Dissolved Organic Compounds on MPs: The Effect of Environmental Factors*

2.3.1. pH

The pH of media is an important factor influencing the adsorption of organic compounds on MPs, due to the fact that variations in pH can change the morphology and surface charge of organic compounds [71]. For instance, PE, PS, and PP have a net positive charge at pH values below their respective points of zero change, i.e., 6.63, 6.69, and 6.76 [54]. The acidic condition favors the partitioning of organic compounds onto MP particles [71]. A few scholars demonstrated a gradual decrease in the adsorption of organic compounds on MPs with increasing pH [54,72,73]. However, studies also showed that the adsorption process of some organic compounds on MPs did not significantly change when the pH was altered [40,48]. For instance, Zhang et al. (2020) proposed that the pH did not significantly affect the adsorption of 9-Nitroanthrene (9NAnt) on PE. This phenomenon can be explained by the sorption of 9NAnt on PE due to the combined effects of hydrophobic interactions and van der Waals forces rather than electrostatic interactions, which are more easily affected by the pH of the medium [48,74].

2.3.2. Ionic Strength

The adsorption of dissolved organic compounds on MPs can be affected by the ionic strength level. An increase in ionic strength (including Cl^-, Ca^{2+}, Na^+, and K^+ ions) could reduce the adsorption of hydrophilic compounds but increase that of hydrophobic organic compounds on MPs [24,40,75–77]. The presence of salt ions could reduce the solubility of organic compounds and promote their hydrophobic adsorption on MPs, but the adsorption enhancement may reach a plateau when the compound's hydrophobicity no longer changes with increasing ionic strength [40]. Guo et al. (2018) found that the presence of potassium ions (K^+) in the solution can cause competition between K^+ ions and DOM for MP adsorption sites. This competition increases with the ionic strength of the solution, which results in a decrease in adsorption [78]. However, Tang et al. (2019) found that the sorption isotherms of benzene on HA were not affected by the concentration of $CaCl_2$ in the solution, which may be due to the fact that the sorption of benzene on HA does not occur via electrostatic interactions or ion exchange [76].

2.3.3. UV Radiation

UV radiation is crucial in the degradation of microplastic particles in the environment [19,79,80]. UV radiation leads to the formation of carbonyl groups in the surface layer of MPs due to the introduction of oxygen. As a result, the polarity of MPs increases, which can lead to a decrease in the adsorption capacity of MPs for nonpolar organic compounds, such as benzene, toluene, ethyl benzene, and xylene [49]. Hüffer et al. (2018) suggested that UV-induced surface functionalization decreases the adsorption coefficient of organic compounds on PS [80]. The adsorption of hydrophilic organic compounds on artificially aged (UV-accelerated aging) MPs was studied by Liu et al. (2019). Their results showed that the adsorption capacities of dissolved organic compounds on aged PS and PVC were much higher than on pristine MPs. The higher adsorption capacity of the aged MPs was caused by the enhanced hydrogen bonding and electrostatic interactions after UV radiation [68].

3. Release of Dissolved Organic Compounds from Microplastics

Although durability and inertness are important characteristics of plastics, studies have shown that plastics can leach additives, such as plasticizers and colorants, when encountering different environmental conditions [81–85]. Mechanical, chemical, and biological processes can cause fragmentation, embrittlement, and modification of MPs in aquatic environments. For instance, Shi et al. (2021) found that the photoaging of polycarbonate MPs could facilitate the fragmentation of PC MPs and enhance bisphenol A release [82]. Romera-Castillo et al. (2018) estimated that, globally, up to 23,600 metric tons of dissolved organic carbon (DOC) can be leached from marine plastics each year. Leached DOC is crucial for the composition and activity of microbe communities in seawater [23]. The environmental modification and weathering processes of MPs in aquatic systems can result in the loss of their surfactants, accelerating the release of organic compounds from MPs into the environment [17].

Besides leaching artificial additives (i.e., plasticizers, colorants, and flame retardants), MPs can also leach absorbed organic compounds into the environment. After irradiation with UV and visible light, Chen et al. (2019) detected the release of organotin compounds (OTCs) from PVC microplastics (MPs) [86]. Four types of OTC, namely dimethyltin (DMT), monomethyltin (MMT), dibutyltin (DBT), and monobutyltin (MBT), were observed to be released from PVC in the dark. However, upon exposure to UV–visible light, only DMT and DBT were detected. This can be attributed to the rapid photodegradation of MMT and MBT [83].

3.1. Effects of MPs' Characteristics

The concentration of organic compounds released from PE is almost four times higher than that released from PP, indicating that the leaching concentration of porous organic matter varies with the type and characteristics of the polymer [23]. Lee et al. (2020) found that the desorption of the MP-derived additive BPA from rubbery PE and PP is usually greater than that from glassy PVC and PS [17]. Research has demonstrated that the molecular chain segments in the glassy subcomponent are denser and more tightly linked than those in the rubbery subcomponent. As a result, the molecules located in the amorphous regions are less densely packed and more prone to decomposition than those in the crystalline regions [87]. The adsorption process in the glass subcomponent is affected by partitioning and space filling [88]. Furthermore, space filling results in a lag in release, primarily due to the conformational changes and distinct physical formation of pores during adsorption and release [89]. It is evident that the release mechanism of MPs can be affected by their distinct morphologies, which alter their conformation.

Organic compounds' release efficiency can be affected by plastic materials such as polymers and additives. Lee et al. (2020) compared the concentration and fluorescence characteristics of DOM released from four different plastic materials (two polymers, PVC and PS, and two additives, DEHP and BPA) under dark and light conditions. Their results showed that even small amounts of additives added to the polymers can increase the possibility of organic compounds being leached from MPs into the aqueous environment [17]. Based on linear model fitting, Zuo et al. (2019) found that the adsorption capacity for phenanthrene (PHEN) was much higher on the biodegradable plastic poly(butylene adipate co-terephthalate) (PBAT) than on conventional plastics, such as PE and PS. The sorption and desorption capacities of MPs are highly dependent on their molecular properties, such as the abundance of rubbery subfractions [90].

3.2. Effects of Environmental Factors

Various environmental factors can affect the release of organic compounds from MPs. Liu et al. (2019) found that the release of the MP additive BPA increased significantly as the pH increased from 3.0 to 11.0 [46]. This result can be explained by the fact that under alkaline conditions, BPA becomes ionized and promotes MP hydrolysis, which increases the solubility and release of BPA [46]. Suhrhoff and Scholz-Böttcher (2015) investigated the impact of salinity, UV radiation, and turbulence on the leaching of plastic-derived organic additives (citrate, phthalates, BPA, etc.). A positive relationship between turbulence and the magnitude of plastic-derived additive leaching was found. In contrast to turbulence, salinity has a minor impact on organic additive leaching from MPs. The effect of salinity highly depends on the inherent properties of plastics. The reason for this is that changes in salinity can alter the water and compound chemistry, affecting the pH, ionic strength, polarity, and other parameters [19]. Chen et al. (2019) found that the release of organotin compounds from PVC particles was inhibited under high-salinity conditions, probably due to organotin's re-adsorption on PVC [86].

The weathering or aging of MPs can lead to changes in MPs' surface and their release of organic compounds. Lee et al. (2020) investigated the leaching of DOM from additive-free MPs in artificial freshwater under UV radiation and dark conditions. Their results showed that UV radiation facilitated the release of DOM from plastic polymers, and the amount of leached DOM was approximately 3% of the total plastic mass [91]. Lee et al. (2020) explored the fluorescence signature of DOM leached from two plastic polymers (PVC and PS), two additives (diethylhexyl (DEHP) and BPA), and two commercial plastics. They found that UV radiation facilitated the leaching of plastic-derived DOM. Under UV radiation, one humic-like component (Ex/Em = 235(290)/410 nm) and one protein/phenol-like fluorescent component (Ex/Em = 270/309 nm) were found to have strong correlations with the polymer-derived DOM [17].

4. Impacts of MPs on Natural DOM in Aquatic Ecosystems

DOM is ubiquitous and plays an essential role in aquatic ecosystems by regulating underwater light and providing an energy source for microorganisms [92,93]. The primary source of autochthonous natural DOM in water is living phytoplankton, which release DOM through lysis, senescence, and grazing [94]. Numerous studies have investigated the toxicity effects of MPs on phytoplankton in water; however, very little is known about the impact of MPs on the phytoplankton production of natural DOM. One recent study by our research group found that light-aged MPs could decrease algal DOM production by 38% and modify the chemical composition of natural DOM by increasing the aromaticity and molecular weight [95]. Compared to direct DOM leaching, the impact of MPs on phytoplankton production of DOM may be more important to natural DOM cycling in water (Figure 1).

Since natural DOM is a heterogeneous organic mixture (ranging from small amino acids to large humic substances) with various functional groups [96], the aforementioned mechanisms (hydrophobic interactions, van der Waals interactions, π–π interactions, hydrogen interactions, and electrostatic interactions) can apply to the interactions between MPs and natural DOM in aquatic environments. In addition, as the molecular structure of natural DOM is much more complex than that of a single organic compound, several mechanisms often coexist for the adsorption of natural DOM on MPs.

Firstly, MPs can affect the physicochemical state of natural DOM in water. Natural DOM in the water column can quickly form organic aggregates due to adsorption on MPs via covalent bonding, hydrogen bonding, or other reactive functional groups, which can lead to the co-precipitation of natural DOM with particulate organic matter (POM). Galgani et al. (2018) observed an accumulation of chromophoric DOM (CDOM) in the sea surface microlayer when MPs were present [97]. Chen et al. (2018) found that 10 ppb nanoplastic particles in the water column accelerated DOM–POM aggregation due to the hydrophobic kinetic assembly [38].

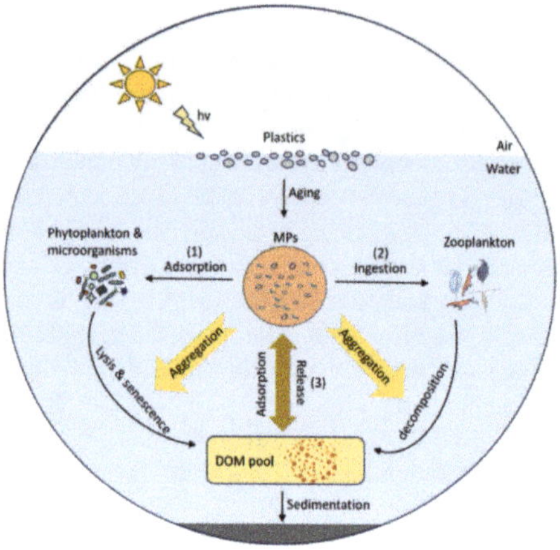

Figure 1. Schematic diagram of the impacts of microplastics (MPs) on the production and transformation of natural dissolved organic matter (DOM) in water. These processes are as follows: (1) MPs affect the autochthonous production of DOM by interacting with phytoplankton and microorganisms; (2) MPs affect DOM production by interacting with zooplankton; and (3) MPs affect DOM production and transformation by releasing and adsorbing organic molecules.

Secondly, MPs can influence the natural DOM cycle by altering the microorganism community in water [98]. Microorganisms are closely associated with MPs in water and sediments and play a crucial role in aquatic biogeochemical cycles [99]. MPs can move up the food chain and quickly become part of the biogeochemical cycles in the water column. They can act as a carbon source for elemental cycling [100]. Planktonic microorganisms can attach to or aggregate with MPs or actively consume or break down MPs [101]. Environmental factors, such as photo-oxidation, can cause microplastics to age and change their surface morphology, roughness, and chemical properties. These changes can increase microbe adhesion, provide favorable conditions for biofilm formation, and ultimately accelerate the biodegradation of MPs [99]. Studies have shown that microorganisms can actively interact with or passively attach to MPs, using available electron acceptors to break down these polymers. These intermediate degradation products may act as electron donors for microbial utilization [101]. On the other hand, the adsorption and release of DOM by MPs can impact the productivity and structure of microbial communities in natural water bodies [101]. Andres et al. (2019) proposed that DOM released by MPs could explain the different characteristics of carbon substrate utilization by microorganisms attached to MPs compared to microorganisms in the surrounding water [102]. Pinto et al. (2020) demonstrated that certain prokaryotes can survive solely on MPs and are relatively abundant in various water masses of the global ocean [103]. Recent studies have shown that the release of DOM from MPs can stimulate microbial utilization of DOC in water [104,105]. However, the article by Oberbeckmann and Labrenz (2020) reviewed the role of microbial interactions with MPs in marine ecosystems and demonstrated that MPs in the ocean represent recalcitrant substances for microorganisms that probably would not be microbially degraded [105].

Finally, MPs can influence natural DOM cycling by interacting with zooplankton. Fecal pellets excreted by zooplankton are not only a food source for marine organisms but also part of the biological pump that contributes to the marine vertical flux of POM, an important source of DOM in water. Cole et al. (2016) found that zooplankton that ingested marine MPs could reduce the density, structural integrity, and sinking rate of their fecal pellets. This interference can hinder the biological transport of carbon from the ocean surface to the deep sea [106]. This result was also observed by Corsi et al. (2020), who demonstrated that the adsorption of MP particles on the surface or their internalization in feces reduced their motility and affected their sinking or floating behavior, ultimately affecting the "sink" of natural DOM [107].

5. Conclusions and Future Perspectives

The present review on the interactions between MPs and dissolved organic compounds addressed the following research issues: the adsorption mechanism of dissolved organic compounds on MPs and environmental factors; the release of dissolved organic compounds from MPs and environmental factors; and the impact of MPs on natural DOM cycling in aquatic ecosystems. Although researchers have made progress in understanding how organic compounds are absorbed and released by MPs in water, the mechanisms that drive the adsorption and release behaviors of dissolved organic compounds on MPs are still not fully understood. More importantly, there has been little investigation into the direct interactions between MPs and natural DOM in aquatic environments. Specifically, the impacts of MPs on the production and transformation of natural DOM in aquatic ecosystems are unanswered questions. In addition, the interactions between aquatic organisms and MPs are numerous and complex, and the nature of their interactions has not yet been thoroughly investigated. The following knowledge gaps require further studies:

(1) Thus far, studies have concentrated on the adsorption of specific organic pollutants on MPs, and little information is available on the direct interactions between MPs and natural sources of dissolved organic compounds.

(2) A few recent studies reported the magnitude of DOM released from MPs in water; however, the significance and contribution of DOM released by MPs to aquatic carbon cycling require further evaluation.

(3) The topic of "MPs have the potential to become a local hotspot for microbial activity and influence the carbon cycling process in water" has been widely proposed. Nevertheless, to what extent MPs influence microorganism metabolism linked to carbon cycling remains an enigma.

This review highlighted the interactions between MPs and natural DOM, offering a limited understanding of the interactions between MPs and dissolved organic compounds in water. Due to the wide variety of MPs and natural DOM, future advances in understanding the impact of MPs on the magnitude and transformation of dissolved organic compounds are urgently needed.

Author Contributions: Conceptualization, W.Y., Q.W. and Y.X.; methodology, Y.X., W.Y., Q.W. and Y.G.; investigation, Y.X., W.Y. and W.Y.; writing—original draft preparation, Q.W.; writing—review and editing, Y.X., Y.G., Q.W., M.X., H.L., Y.Z. and C.L.; supervision, Y.X. and C.L.; funding acquisition, Y.X. and C.L. All authors have read and agreed to the published version of the manuscript.

Funding: This research was jointly funded by the National Key Research and Development Program (No. 2021YFC3201004), the Taishan Scholar Foundation of Shandong Province (No. tsqn201909126), the National Natural Science Foundation of China (No. 42103029), the Natural Science Foundation of Shandong Province (No. ZR2021MD125), and the Foundation of Key Laboratory of Yangtze River Water Environment, Ministry of Education (Tongji University), China (No. YRWEF202102).

Data Availability Statement: Not applicable.

Conflicts of Interest: The authors declare no conflict of interest.

References

1. Li, C.; Busquets, R.; Campos, L.C. Assessment of Microplastics in Freshwater Systems. *Sci. Total Environ.* **2020**, *707*, 135578. [CrossRef]
2. Sun, Y.; Ji, J.; Tao, J.; Yang, Y.; Wu, D.; Han, L.; Li, S.; Wang, J. Current Advances in Interactions between Microplastics and Dissolved Organic Matters in Aquatic and Terrestrial Ecosystems. *TrAC—Trends Anal. Chem.* **2023**, *158*, 116882. [CrossRef]
3. Sheridan, E.A.; Fonvielle, J.A.; Cottingham, S.; Zhang, Y.; Dittmar, T.; Aldridge, D.C.; Tanentzap, A.J. Plastic Pollution Fosters More Microbial Growth in Lakes than Natural Organic Matter. *Nat. Commun.* **2022**, *13*, 4175. [CrossRef] [PubMed]

4. Eriksen, M.; Lebreton, L.C.M.; Carson, H.S.; Thiel, M.; Moore, C.J.; Borerro, J.C.; Galgani, F.; Ryan, P.G.; Reisser, J. Plastic Pollution in the World's Oceans: More than 5 Trillion Plastic Pieces Weighing over 250,000 Tons Afloat at Sea. *PLoS ONE* **2014**, *9*, e111913. [CrossRef] [PubMed]
5. Chen, C.; Yu, G.; Wang, B.; Li, F.; Liu, H.; Zhang, W. Lifetime Prediction of Non-Woven Face Masks in Ocean and Contributions to Microplastics and Dissolved Organic Carbon. *J. Hazard. Mater.* **2022**, *441*, 129816. [CrossRef]
6. Shiu, R.-F.; Vazquez, C.I.; Tsai, Y.Y.; Torres, G.V.; Chen, C.S.; Santschi, P.H.; Quigg, A.; Chin, W.C. Nano-Plastics Induce Aquatic Particulate Organic Matter (Microgels) Formation. *Sci. Total Environ.* **2020**, *706*, 135681. [CrossRef]
7. Cowger, W.C.; Gray, A.B.; Eriksen, M.; Moore, C.; Thiel, M. Evaluating Wastewater Effluent as a Source of Microplastics in Environmental Samples. In *Microplastics in Water and Wastewater*; Karapanagioti, H.K., Kalavrouziotis, I.K., Eds.; IWA Publishing: London, UK, 2019; ISBN 9781789060034.
8. Brandon, J.A.; Jones, W.; Ohman, M.D. Multidecadal Increase in Plastic Particles in Coastal Ocean Sediments. *Sci. Adv.* **2019**, *5*, eaax0587. [CrossRef]
9. Al-Thawadi, S. Microplastics and Nanoplastics in Aquatic Environments: Challenges and Threats to Aquatic Organisms. *Arab. J. Sci. Eng.* **2020**, *45*, 4419–4440. [CrossRef]
10. Hanachi, P.; Khoshnamvand, M.; Walker, T.R.; Hamidian, A.H. Nano-Sized Polystyrene Plastics Toxicity to Microalgae Chlorella Vulgaris: Toxicity Mitigation Using Humic Acid. *Aquat. Toxicol.* **2022**, *245*, 106123. [CrossRef]
11. Shiu, R.-F.; Vazquez, C.I.; Chiang, C.-Y.; Chiu, M.-H.; Chen, C.-S.; Ni, C.-W.; Gong, G.-C.; Quigg, A.; Santschi, P.H.; Chin, W.-C. Nano- and Microplastics Trigger Secretion of Protein-Rich Extracellular Polymeric Substances from Phytoplankton. *Sci. Total Environ.* **2020**, *748*, 141469. [CrossRef]
12. California SB-1422 California Safe Drinking Water Act: Microplastics. 2018. Volume 902. p. 92. Available online: https://www.cawrecycles.org/sb-1422-portantino (accessed on 28 September 2018).
13. Carpenter, E.J.; Anderson, S.J.; Harvey, G.R.; Miklas, H.P.; Peck, B.B. Polystyrene Spherules in Coastal Waters. *Science* **1972**, *178*, 749–750. [CrossRef]
14. Kalogerakis, N.; Karkanorachaki, K.; Kalogerakis, G.C.; Triantafyllidi, E.I.; Gotsis, A.D.; Partsinevelos, P.; Fava, F. Microplastics Generation: Onset of Fragmentation of Polyethylene Films in Marine Environment Mesocosms. *Front. Mar. Sci.* **2017**, *4*, 84. [CrossRef]
15. Andrady, A.L. Microplastics in the Marine Environment. *Mar. Pollut. Bull.* **2011**, *62*, 1596–1605. [CrossRef] [PubMed]
16. Chae, Y.; An, Y.-J. Current Research Trends on Plastic Pollution and Ecological Impacts on the Soil Ecosystem: A Review. *Environ. Pollut.* **2018**, *240*, 387–395. [CrossRef] [PubMed]
17. Lee, Y.K.; Murphy, K.R.; Hur, J. Fluorescence Signatures of Dissolved Organic Matter Leached from Microplastics: Polymers and Additives. *Environ. Sci. Technol.* **2020**, *54*, 11905–11914. [CrossRef]
18. Reimonn, G.; Lu, T.; Gandhi, N.; Chen, W.-T.T. Review of Microplastic Pollution in the Environment and Emerging Recycling Solutions. *J. Renew. Mater.* **2019**, *7*, 1251–1268. [CrossRef]
19. Suhrhoff, T.J.; Scholz-Böttcher, B.M. Qualitative Impact of Salinity, UV Radiation and Turbulence on Leaching of Organic Plastic Additives from Four Common Plastics—A Lab Experiment. *Mar. Pollut. Bull.* **2016**, *102*, 84–94. [CrossRef] [PubMed]
20. Hermabessiere, L.; Dehaut, A.; Paul-Pont, I.; Lacroix, C.; Jezequel, R.; Soudant, P.; Duflos, G. Occurrence and Effects of Plastic Additives on Marine Environments and Organisms: A Review. *Chemosphere* **2017**, *182*, 781–793. [CrossRef]
21. Lithner, D.; Larsson, A.; Dave, G. Environmental and Health Hazard Ranking and Assessment of Plastic Polymers Based on Chemical Composition. *Sci. Total Environ.* **2011**, *409*, 3309–3324. [CrossRef]
22. Lithner, D.; Damberg, J.; Dave, G.; Larsson, Å. Leachates from Plastic Consumer Products—Screening for Toxicity with Daphnia Magna. *Chemosphere* **2009**, *74*, 1195–1200. [CrossRef]
23. Romera-Castillo, C.; Pinto, M.; Langer, T.M.; Álvarez-Salgado, X.A.; Herndl, G.J. Dissolved Organic Carbon Leaching from Plastics Stimulates Microbial Activity in the Ocean. *Nat. Commun.* **2018**, *9*, 1430. [CrossRef] [PubMed]
24. Velzeboer, I.; Kwadijk, C.J.A.F.; Koelmans, A.A. Strong Sorption of PCBs to Nanoplastics, Microplastics, Carbon Nanotubes, and Fullerenes. *Environ. Sci. Technol.* **2014**, *48*, 4869–4876. [CrossRef]
25. Ziajahromi, S.; Kumar, A.; Neale, P.A.; Leusch, F.D.L. Effects of Polyethylene Microplastics on the Acute Toxicity of a Synthetic Pyrethroid to Midge Larvae (*Chironomus tepperi*) in Synthetic and River Water. *Sci. Total Environ.* **2019**, *671*, 971–975. [CrossRef]
26. Bakir, A.; O'Connor, I.A.; Rowland, S.J.; Hendriks, A.J.; Thompson, R.C. Relative Importance of Microplastics as a Pathway for the Transfer of Hydrophobic Organic Chemicals to Marine Life. *Environ. Pollut.* **2016**, *219*, 56–65. [CrossRef]
27. Lee, H.; Shim, W.J.; Kwon, J.H. Sorption Capacity of Plastic Debris for Hydrophobic Organic Chemicals. *Sci. Total Environ.* **2014**, *470–471*, 1545–1552. [CrossRef] [PubMed]
28. Yu, F.; Yang, C.; Zhu, Z.; Bai, X.; Ma, J. Adsorption Behavior of Organic Pollutants and Metals on Micro/Nanoplastics in the Aquatic Environment. *Sci. Total Environ.* **2019**, *694*, 133643. [CrossRef]
29. Ryan, P.G.; Connell, A.D.; Gardner, B.D. Plastic Ingestion and PCBs in Seabirds: Is There a Relationship? *Mar. Pollut. Bull.* **1988**, *19*, 174–176. [CrossRef]
30. Tanaka, K.; Takada, H.; Yamashita, R.; Mizukawa, K.; Fukuwaka, M.; Watanuki, Y. Accumulation of Plastic-Derived Chemicals in Tissues of Seabirds Ingesting Marine Plastics. *Mar. Pollut. Bull.* **2013**, *69*, 219–222. [CrossRef]

31. Potthoff, A.; Oelschlägel, K.; Schmitt-Jansen, M.; Rummel, C.D.; Kühnel, D. From the Sea to the Laboratory: Characterization of Microplastic as Prerequisite for the Assessment of Ecotoxicological Impact. *Integr. Environ. Assess. Manag.* **2017**, *13*, 500–504. [CrossRef]
32. Engler, R.E. The Complex Interaction between Marine Debris and Toxic Chemicals in the Ocean. *Environ. Sci. Technol.* **2012**, *46*, 12302–12315. [CrossRef]
33. Galgani, L.; Loiselle, S.A. Plastic Pollution Impacts on Marine Carbon Biogeochemistry. *Environ. Pollut.* **2021**, *268*, 115598. [CrossRef] [PubMed]
34. Bolan, N.S.; Adriano, D.C.; Kunhikrishnan, A.; James, T.; McDowell, R.; Senesi, N. *Dissolved Organic Matter: Biogeochemistry, Dynamics, and Environmental Significance in Soils*, 1st ed.; Elsevier Inc.: Amsterdam, The Netherlands, 2011; Volume 110, ISBN 9780123855312.
35. Ma, Y.; Huang, A.; Cao, S.; Sun, F.; Wang, L.; Guo, H.; Ji, R. Effects of Nanoplastics and Microplastics on Toxicity, Bioaccumulation, and Environmental Fate of Phenanthrene in Fresh Water. *Environ. Pollut.* **2016**, *219*, 166–173. [CrossRef] [PubMed]
36. Yonkos, L.T.; Friedel, E.A.; Perez-Reyes, A.C.; Ghosal, S.; Arthur, C.D. Microplastics in Four Estuarine Rivers in the Chesapeake Bay, U.S.A. *Environ. Sci. Technol.* **2014**, *48*, 14195–14202. [CrossRef]
37. Mei, W.; Chen, G.; Bao, J.; Song, M.; Li, Y.; Luo, C. Interactions between Microplastics and Organic Compounds in Aquatic Environments: A Mini Review. *Sci. Total Environ.* **2020**, *736*, 139472. [CrossRef] [PubMed]
38. Chen, C.-S.; Le, C.; Chiu, M.-H.; Chin, W.-C. The Impact of Nanoplastics on Marine Dissolved Organic Matter Assembly. *Sci. Total Environ.* **2018**, *634*, 316–320. [CrossRef] [PubMed]
39. Abdurahman, A.; Cui, K.; Wu, J.; Li, S.; Gao, R.; Dai, J.; Liang, W.; Zeng, F. Adsorption of Dissolved Organic Matter (DOM) on Polystyrene Microplastics in Aquatic Environments: Kinetic, Isotherm and Site Energy Distribution Analysis. *Ecotoxicol. Environ. Saf.* **2020**, *198*, 110658. [CrossRef] [PubMed]
40. Fang, S.; Yu, W.; Li, C.; Liu, Y.; Qiu, J.; Kong, F. Adsorption Behavior of Three Triazole Fungicides on Polystyrene Microplastics. *Sci. Total Environ.* **2019**, *691*, 1119–1126. [CrossRef]
41. Guo, X.; Wang, J. Sorption of Antibiotics onto Aged Microplastics in Freshwater and Seawater. *Mar. Pollut. Bull.* **2019**, *149*, 110511. [CrossRef]
42. Liu, F.; Liu, G.; Zhu, Z.; Wang, S.; Zhao, F. Interactions between Microplastics and Phthalate Esters as Affected by Microplastics Characteristics and Solution Chemistry. *Chemosphere* **2019**, *214*, 688–694. [CrossRef]
43. Tourinho, P.S.; Kočí, V.; Loureiro, S.; van Gestel, C.A.M. Partitioning of Chemical Contaminants to Microplastics: Sorption Mechanisms, Environmental Distribution and Effects on Toxicity and Bioaccumulation. *Environ. Pollut.* **2019**, *252*, 1246–1256. [CrossRef]
44. Zhang, Y.; Liu, X.; Wang, M.; Qin, B. Compositional Differences of Chromophoric Dissolved Organic Matter Derived from Phytoplankton and Macrophytes. *Org. Geochem.* **2013**, *55*, 26–37. [CrossRef]
45. Liu, J.; Wang, Y.; Jiang, B.; Wang, L.; Chen, J.; Guo, H.; Ji, R. Degradation, Metabolism, and Bound-Residue Formation and Release of Tetrabromobisphenol A in Soil during Sequential Anoxic-Oxic Incubation. *Environ. Sci. Technol.* **2013**, *47*, 8348–8354. [CrossRef]
46. Liu, X.; Shi, H.; Xie, B.; Dionysiou, D.D.; Zhao, Y. Microplastics as Both a Sink and a Source of Bisphenol A in the Marine Environment. *Environ. Sci. Technol.* **2019**, *53*, 10188–10196. [CrossRef] [PubMed]
47. Holmes, L.A.; Turner, A.; Thompson, R.C. Adsorption of Trace Metals to Plastic Resin Pellets in the Marine Environment. *Environ. Pollut.* **2012**, *160*, 42–48. [CrossRef]
48. Zhang, J.; Chen, H.; He, H.; Cheng, X.; Ma, T.; Hu, J.; Yang, S.; Li, S.; Zhang, L. Adsorption Behavior and Mechanism of 9-Nitroanthracene on Typical Microplastics in Aqueous Solutions. *Chemosphere* **2020**, *245*, 125628. [CrossRef] [PubMed]
49. Müller, A.; Becker, R.; Dorgerloh, U.; Simon, F.G.; Braun, U. The Effect of Polymer Aging on the Uptake of Fuel Aromatics and Ethers by Microplastics. *Environ. Pollut.* **2018**, *240*, 639–646. [CrossRef]
50. Liu, X.; Zheng, M.; Wang, L.; Ke, R.; Lou, Y.; Zhang, X.; Dong, X.; Zhang, Y. Sorption Behaviors of Tris-(2,3-Dibromopropyl) Isocyanurate and Hexabromocyclododecanes on Polypropylene Microplastics. *Mar. Pollut. Bull.* **2018**, *135*, 581–586. [CrossRef]
51. Xiong, W.; Zeng, G.; Yang, Z.; Zhou, Y.; Zhang, C.; Cheng, M.; Liu, Y.; Hu, L.; Wan, J.; Zhou, C.; et al. Adsorption of Tetracycline Antibiotics from Aqueous Solutions on Nanocomposite Multi-Walled Carbon Nanotube Functionalized MIL-53(Fe) as New Adsorbent. *Sci. Total Environ.* **2018**, *627*, 235–244. [CrossRef]
52. Chen, W.; Ouyang, Z.Y.; Qian, C.; Yu, H.Q. Induced Structural Changes of Humic Acid by Exposure of Polystyrene Microplastics: A Spectroscopic Insight. *Environ. Pollut.* **2018**, *233*, 1–7. [CrossRef]
53. Zhang, L.; Luo, L.; Zhang, S. Integrated Investigations on the Adsorption Mechanisms of Fulvic and Humic Acids on Three Clay Minerals. *Colloids Surf. A Physicochem. Eng. Asp.* **2012**, *406*, 84–90. [CrossRef]
54. Li, J.; Zhang, K.; Zhang, H. Adsorption of Antibiotics on Microplastics. *Environ. Pollut.* **2018**, *237*, 460–467. [CrossRef]
55. Xu, B.; Liu, F.; Brookes, P.C.; Xu, J. Microplastics Play a Minor Role in Tetracycline Sorption in the Presence of Dissolved Organic Matter. *Environ. Pollut.* **2018**, *240*, 87–94. [CrossRef] [PubMed]
56. Guo, X.; Wang, X.; Zhou, X.; Kong, X.; Tao, S.; Xing, B. Sorption of Four Hydrophobic Organic Compounds by Three Chemically Distinct Polymers: Role of Chemical and Physical Composition. *Environ. Sci. Technol.* **2012**, *46*, 7252–7259. [CrossRef]
57. Wu, X.; Liu, P.; Huang, H.; Gao, S. Adsorption of Triclosan onto Different Aged Polypropylene Microplastics: Critical Effect of Cations. *Sci. Total Environ.* **2020**, *717*, 137033. [CrossRef] [PubMed]
58. Hüffer, T.; Hofmann, T. Sorption of Non-Polar Organic Compounds by Micro-Sized Plastic Particles in Aqueous Solution. *Environ. Pollut.* **2016**, *214*, 194–201. [CrossRef] [PubMed]

59. Li, R.; Tan, H.; Zhang, L.; Wang, S.; Wang, Y.; Yu, K. The Implications of Water Extractable Organic Matter (WEOM) on the Sorption of Typical Parent, Alkyl and N/O/S-Containing Polycyclic Aromatic Hydrocarbons (PAHs) by Microplastics. *Ecotoxicol. Environ. Saf.* **2018**, *156*, 176–182. [CrossRef]
60. Liu, P.; Qian, L.; Wang, H.; Zhan, X.; Lu, K.; Gu, C.; Gao, S. New Insights into the Aging Behavior of Microplastics Accelerated by Advanced Oxidation Processes. *Environ. Sci. Technol.* **2019**, *53*, 3579–3588. [CrossRef] [PubMed]
61. Charles, E.; Carraher, J. *Polymer Chemistry: Polymer Density*, 9th ed.; CRC Press: Boca Raton, FL, USA; Taylor & Francis Group: Boca Raton, FL, USA, 2013; ISBN 9781466552203.
62. Fred-Ahmadu, O.H.; Bhagwat, G.; Oluyoye, I.; Benson, N.U.; Ayejuyo, O.O.; Palanisami, T. Interaction of Chemical Contaminants with Microplastics: Principles and Perspectives. *Sci. Total Environ.* **2020**, *706*, 135978. [CrossRef] [PubMed]
63. Wang, J.; Liu, X.; Liu, G.; Zhang, Z.; Wu, H.; Cui, B.; Bai, J.; Zhang, W. Size Effect of Polystyrene Microplastics on Sorption of Phenanthrene and Nitrobenzene. *Ecotoxicol. Environ. Saf.* **2019**, *173*, 331–338. [CrossRef] [PubMed]
64. Hu, J.Q.; Yang, S.Z.; Guo, L.; Xu, X.; Yao, T.; Xie, F. Microscopic Investigation on the Adsorption of Lubrication Oil on Microplastics. *J. Mol. Liq.* **2017**, *227*, 351–355. [CrossRef]
65. Liu, X.; Xu, J.; Zhao, Y.; Shi, H.; Huang, C.H. Hydrophobic Sorption Behaviors of 17B-Estradiol on Environmental Microplastics. *Chemosphere* **2019**, *226*, 726–735. [CrossRef]
66. Enders, K.; Lenz, R.; Stedmon, C.A.; Nielsen, T.G. Abundance, Size and Polymer Composition of Marine Microplastics $\geq$ 10 Mm in the Atlantic Ocean and Their Modelled Vertical Distribution. *Mar. Pollut. Bull.* **2015**, *100*, 70–81. [CrossRef]
67. Zhao, L.; Rong, L.; Xu, J.; Lian, J.; Wang, L.; Sun, H. Sorption of Five Organic Compounds by Polar and Nonpolar Microplastics. *Chemosphere* **2020**, *257*, 127206. [CrossRef] [PubMed]
68. Liu, X.; Ge, W.; Zhang, X.; Chai, C.; Wu, J.; Xiang, D.; Chen, X. Biodegradation of Aged Polycyclic Aromatic Hydrocarbons in Agricultural Soil by Paracoccus Sp. LXC Combined with Humic Acid and Spent Mushroom Substrate. *J. Hazard. Mater.* **2019**, *379*, 120820. [CrossRef] [PubMed]
69. Hirsch, S.G.; Barel, B.; Segal, E. Characterization of Surface Phenomena: Probing Early Stage Degradation of Low-Density Polyethylene Films. *Polym. Eng. Sci.* **2019**, *59*, E129–E137. [CrossRef]
70. Fries, E.; Zarfl, C. Sorption of Polycyclic Aromatic Hydrocarbons (PAHs) to Low and High Density Polyethylene (PE). *Environ. Sci. Pollut. Res.* **2012**, *19*, 1296–1304. [CrossRef] [PubMed]
71. Elizalde-Velázquez, A.; Subbiah, S.; Anderson, T.A.; Green, M.J.; Zhao, X.; Cañas-Carrell, J.E. Sorption of Three Common Nonsteroidal Anti-Inflammatory Drugs (NSAIDs) to Microplastics. *Sci. Total Environ.* **2020**, *715*, 136974. [CrossRef]
72. Li, J.; Liu, H.; Paul Chen, J. Microplastics in Freshwater Systems: A Review on Occurrence, Environmental Effects, and Methods for Microplastics Detection. *Water Res.* **2018**, *137*, 362–374. [CrossRef]
73. Liang, L.; Luo, L.; Zhang, S. Adsorption and Desorption of Humic and Fulvic Acids on SiO2 Particles at Nano- and Micro-Scales. *Colloids Surf. A Physicochem. Eng. Asp.* **2011**, *384*, 126–130. [CrossRef]
74. Llorca, M.; Schirinzi, G.; Martínez, M.; Barceló, D.; Farré, M. Adsorption of Perfluoroalkyl Substances on Microplastics under Environmental Conditions. *Environ. Pollut.* **2018**, *235*, 680–691. [CrossRef]
75. Duan, L.; Liu, J.; Xin, Y.; Larssen, T. Air-Pollution Emission Control in China: Impacts on Soil Acidification Recovery and Constraints Due to Drought. *Sci. Total Environ.* **2013**, *463–464*, 1031–1041. [CrossRef] [PubMed]
76. Tang, Z.; Li, Y.; Yang, Z.; Liu, D.; Tang, M.; Yang, S.; Tang, Y. Characteristic and Mechanism of Sorption and Desorption of Benzene on Humic Acid. *Environ. Sci. Pollut. Res.* **2019**, *26*, 20277–20285. [CrossRef] [PubMed]
77. Wang, F.; Shih, K.M.; Li, X.Y. The Partition Behavior of Perfluorooctanesulfonate (PFOS) and Perfluorooctanesulfonamide (FOSA) on Microplastics. *Chemosphere* **2015**, *119*, 841–847. [CrossRef] [PubMed]
78. Guo, X.; Pang, J.; Chen, S.; Jia, H. Sorption Properties of Tylosin on Four Different Microplastics. *Chemosphere* **2018**, *209*, 240–245. [CrossRef]
79. Liu, G.; Zhu, Z.; Yang, Y.; Sun, Y.; Yu, F.; Ma, J. Sorption Behavior and Mechanism of Hydrophilic Organic Chemicals to Virgin and Aged Microplastics in Freshwater and Seawater. *Environ. Pollut.* **2019**, *246*, 26–33. [CrossRef]
80. Hüffer, T.; Weniger, A.K.; Hofmann, T. Sorption of Organic Compounds by Aged Polystyrene Microplastic Particles. *Environ. Pollut.* **2018**, *236*, 218–225. [CrossRef]
81. Hahladakis, J.N.; Velis, C.A.; Weber, R.; Iacovidou, E.; Purnell, P. An Overview of Chemical Additives Present in Plastics: Migration, Release, Fate and Environmental Impact during Their Use, Disposal and Recycling. *J. Hazard. Mater.* **2018**, *344*, 179–199. [CrossRef] [PubMed]
82. Shi, Y.; Liu, P.; Wu, X.; Shi, H.; Huang, H.; Wang, H.; Gao, S. Insight into Chain Scission and Release Profiles from Photodegradation of Polycarbonate Microplastics. *Water Res.* **2021**, *195*, 116980. [CrossRef]
83. Chen, C.; Chen, L.; Yao, Y.; Artigas, F.; Huang, Q.; Zhang, W. Organotin Release from Polyvinyl Chloride Microplastics and Concurrent Photodegradation in Water: Impacts from Salinity, Dissolved Organic Matter, and Light Exposure. *Environ. Sci. Technol.* **2019**, *53*, 10741–10752. [CrossRef]
84. Williams, T.; Walsh, C.; Murray, K.; Subir, M. Interactions of Emerging Contaminants with Model Colloidal Microplastics, C60 Fullerene, and Natural Organic Matter-Effect of Surface Functional Group and Adsorbate Properties. *Environ. Sci. Process. Impacts* **2020**, *22*, 1190–1200. [CrossRef]

85. Jiang, M.; Hu, L.; Lu, A.; Liang, G.; Lin, Z.; Zhang, T.; Xu, L.; Li, B.; Gong, W. Strong Sorption of Two Fungicides onto Biodegradable Microplastics with Emphasis on the Negligible Role of Environmental Factors. *Environ. Pollut.* **2020**, *267*, 115496. [CrossRef] [PubMed]
86. Chen, C.; Chen, L.; Xue, R.; Huang, Q.; Wu, L.; Ye, S.; Zhang, W. Spatiotemporal Variation and Source Apportionment of Organotin Compounds in Sediments in the Yangtze Estuary. *Environ. Sci. Eur.* **2019**, *31*, 24. [CrossRef]
87. Tokiwa, Y.; Calabia, B.P.; Ugwu, C.U.; Aiba, S. Biodegradability of Plastics. *Int. J. Mol. Sci.* **2009**, *10*, 3722–3742. [CrossRef]
88. Bakir, A.; Rowland, S.J.; Thompson, R.C. Enhanced Desorption of Persistent Organic Pollutants from Microplastics under Simulated Physiological Conditions. *Environ. Pollut.* **2014**, *185*, 16–23. [CrossRef] [PubMed]
89. Shi, X.; Fu, H.; Li, Y.; Mao, J.; Zheng, S.; Zhu, D. Impact of Coal Structural Heterogeneity on the Nonideal Sorption of Organic Contaminants. *Environ. Toxicol. Chem.* **2011**, *30*, 1310–1319. [CrossRef] [PubMed]
90. Zuo, L.; Li, H.; Lin, L.; Sun, Y.; Diao, Z.; Liu, S.; Zhang, Z.; Xu, X. Sorption and Desorption of Phenanthrene on Biodegradable Poly(Butylene Adipate Co-Terephtalate) Microplastics. *Chemosphere* **2019**, *215*, 25–32. [CrossRef]
91. Lee, Y.K.; Romera-Castillo, C.; Hong, S.; Hur, J. Characteristics of Microplastic Polymer-Derived Dissolved Organic Matter and Its Potential as a Disinfection Byproduct Precursor. *Water Res.* **2020**, *175*, 115678. [CrossRef]
92. Williamson, C.E.; Morris, D.P.; Pace, M.L.; Olson, O.G. Dissolved Organic Carbon and Nutrients as Regulators of Lake Ecosystems: Resurrection of a More Integrated Paradigm. *Limnol. Oceanogr.* **1999**, *44*, 795–803. [CrossRef]
93. Coble, P.G. Marine Optical Biogeochemistry: The Chemistry of Ocean Color. *Chem. Rev.* **2007**, *107*, 402–418. [CrossRef]
94. Moran, M.A.; Ferrer-González, F.X.; Fu, H.; Nowinski, B.; Olofsson, M.; Powers, M.A.; Schreier, J.E.; Schroer, W.F.; Smith, C.B.; Uchimiya, M. The Ocean's Labile DOC Supply Chain. *Limnol. Oceanogr.* **2022**, *67*, 1007–1021. [CrossRef]
95. Xiao, Y.; Wang, Q.; Li, P.; Xu, M.; Zhou, Y.; Li, H.; Yan, W.; Liu, C.; Vähätalo, A.V. Impact of Light-Aged Microplastic on Microalgal Production of Dissolved Organic Matter. *Sci. Total Environ.* **2023**, *889*, 164312. [CrossRef]
96. Wells, M.J.M.; Stretz, H.A. Supramolecular Architectures of Natural Organic Matter. *Sci. Total Environ.* **2019**, *671*, 1125–1133. [CrossRef]
97. Galgani, L.; Engel, A.; Rossi, C.; Donati, A.; Loiselle, S.A. Polystyrene Microplastics Increase Microbial Release of Marine Chromophoric Dissolved Organic Matter in Microcosm Experiments. *Sci. Rep.* **2018**, *8*, 14635. [CrossRef]
98. Hitchcock, J.N. Microplastics Can Alter Phytoplankton Community Composition. *Sci. Total Environ.* **2022**, *819*, 153074. [CrossRef] [PubMed]
99. Zettler, E.R.; Mincer, T.J.; Amaral-Zettler, L.A. Life in the "Plastisphere": Microbial Communities on Plastic Marine Debris. *Environ. Sci. Technol.* **2013**, *47*, 7137–7146. [CrossRef] [PubMed]
100. Romera-Castillo, C.; Mallenco-Fornies, R.; Saá-Yánez, M.; Álvarez-Salgado, X.A. Leaching and Bioavailability of Dissolved Organic Matter from Petrol-Based and Biodegradable Plastics. *Mar. Environ. Res.* **2022**, *176*, 105607. [CrossRef]
101. Rogers, K.L.; Carreres-Calabuig, J.A.; Gorokhova, E.; Posth, N.R. Micro-by-micro Interactions: How Microorganisms Influence the Fate of Marine Microplastics. *Limnol. Oceanogr. Lett.* **2020**, *5*, 18–36. [CrossRef]
102. Arias-Andres, M.; Rojas-Jimenez, K.; Grossart, H.P. Collateral Effects of Microplastic Pollution on Aquatic Microorganisms: An Ecological Perspective. *TrAC—Trends Anal. Chem.* **2019**, *112*, 234–240. [CrossRef]
103. Pinto, M.; Polania Zenner, P.; Langer, T.M.; Harrison, J.; Simon, M.; Varela, M.M.; Herndl, G.J. Putative Degraders of Low-Density Polyethylene-Derived Compounds Are Ubiquitous Members of Plastic-Associated Bacterial Communities in the Marine Environment. *Environ. Microbiol.* **2020**, *22*, 4779–4793. [CrossRef]
104. Chae, Y.; Hong, S.H.; An, Y.J. Photosynthesis Enhancement in Four Marine Microalgal Species Exposed to Expanded Polystyrene Leachate. *Ecotoxicol. Environ. Saf.* **2020**, *189*, 109936. [CrossRef]
105. Oberbeckmann, S.; Labrenz, M. Marine Microbial Assemblages on Microplastics: Diversity, Adaptation, and Role in Degradation. *Ann. Rev. Mar. Sci.* **2020**, *12*, 209–232. [CrossRef] [PubMed]
106. Cole, M.; Lindeque, P.K.; Fileman, E.; Clark, J.; Lewis, C.; Halsband, C.; Galloway, T.S. Microplastics Alter the Properties and Sinking Rates of Zooplankton Faecal Pellets. *Environ. Sci. Technol.* **2016**, *50*, 3239–3246. [CrossRef] [PubMed]
107. Corsi, I.; Bergami, E.; Grassi, G. Behavior and Bio-Interactions of Anthropogenic Particles in Marine Environment for a More Realistic Ecological Risk Assessment. *Front. Environ. Sci.* **2020**, *8*, 60. [CrossRef]

water

MDPI

Article

Slaughterhouse Wastewater Properties Assessment by Modern and Classic Methods

Ramona Crainic [1,2] and Radu Fechete [2,*]

1 Faculty of Physics, Doctoral School, Babeş-Bolyai University, 1 Kogălniceanu, 400084 Cluj-Napoca, Romania; ramona.crainic95@gmail.com
2 Physics and Chemistry Department, Technical University of Cluj-Napoca, Str. Muncii 103-105, 400641 Cluj-Napoca, Romania
* Correspondence: rfechete@phys.utcluj.ro; Tel.: +40-741-770-595

Abstract: Advanced ^{1}H Nuclear Magnetic Resonance (NMR) relaxometry and diffusometry methods and VIS-nearIR spectroscopy combined with pH, electrical conductivity (EC) and totally dissolved solids (TDSSs) measurements were used to assess the properties of wastewater collected from a chicken slaughterhouse in each step of the treatment process (wastewater before treatment, biologically treated wastewater, chemically treated wastewater and discharged wastewater) and from sludge. The ^{1}H NMR Carr–Purcell–Meiboom–Gill (CPMG) and Pulsed-Gradient-Stimulated-Echo (PGSE) decay curves recorded for all samples of wastewater were analyzed by inverse Laplace transform (ILT) to obtain the distributions of transverse relaxation times T_2 and diffusion coefficient D. The VIS-nearIR total absorbance, T_2-values, D-values, pH, EC and TDSS parameters were used for statistical analysis in principal component (PCA). The ^{1}H T_2-distributions measured for the slaughterhouse wastewater lie in two main regions reflecting the number of dissolved solids or the distribution of undissolved solids. The PCA analysis successfully differentiates between polluted and less polluted wastewaters and sludge. The wastewater treatment applied by the slaughterhouse is efficient. The recommended methods for wastewater monitoring are the NMR T_2- and D-distributions and EC, TDSSs and NMR-D diffusion coefficient. Finally, Machine Learning algorithms are used to provide prediction maps of wastewater treatment stage.

Keywords: slaughterhouse wastewater and sludge; ^{1}H NMR relaxometry and diffusometry; VIS-nearIR spectroscopy; pH; EC and TDSSs measurements; PCA (principal component analysis); AI-prediction using machine learning

1. Introduction

Nuclear Magnetic Resonance (NMR) is a powerful analytical tool, extensively used for the characterization of microscopic properties of many classes of materials. However, it is not yet favored in the assessment of wastewater properties, particularly in that collected from slaughterhouses. Classical measurements of wastewater characteristics are mainly based on determining the global physicochemical parameters, such as electrical conductivity (EC), total suspended solids (TSSs), total dissolved solids (TDSSs), apparent color, turbidity, pH, chemical oxygen demand (COD) and ammoniacal nitrogen (NH_3-N) of poultry slaughterhouse wastewater (PSW) [1]. Distilled water has a poor electrical conductivity due to the lack of ions. The presence of electrically charged particles is necessary to support the electrical conduction in solutions. Therefore, measuring high electrical conductivity in wastewater clearly indicates the presence of various types of particles, and thus a certain degree of wastewater pollution [2]. Determining the total suspended solids and the total dissolved solids can indicate the amount of undissolved solid pollutants and that of dissolved pollutants, respectively. The TDSSs and/or the TSSs values are usually proportional to the amounts of these solids in the aqueous sample. Depending on the size and the electrical properties of the particles in the wastewater, liquids can present certain sensitivity to the incident light. Therefore, an apparent color of wastewater can be associated with the presence of specific pollutants. Otherwise, the overall characterization within the visible range can be carried out by performing turbidity measurements. The presence of dissolved or undissolved pollutants may increase its turbidity and therefore reduce the intensity of light transmitted through wastewater samples [3]. Water is a good solvent for a large range of compounds. Therefore, the measurement of wastewater pH value can indicate the pollution level [4,5]. The amount of organic matter in wastewater samples can be determined by classical methods such as: (i) total organic carbon (TOC), directly indicating the organic matter content; (ii) biochemical oxygen demand (BOD), and (iii) chemical oxygen demand measurements. The last two (BOD and COD) offer an indirect quantification of the total organic matter content in aqueous samples. The biochemical oxygen demand is related to the biodegradable organic content of samples. This can be quantified by measuring the oxygen consumption by microorganisms present in the water samples [6,7]. Another indicator of the total organic content is the presence of ammonia, an inorganic compound that demands and consumes oxygen during the process of conversion to nitrate. Other parameters that may complete characterization of wastewater quality may include: (i) the chemical variables such as, ammonia (NH_3), chloride (Cl^-), nitrates ($NO_3{}^-$), orthophosphates ($-PO_4^{-3}$), sulphates ($SO_4{}^{2-}$), oxidation reduction potential (ORP), total nitrogen (TN) or total phosphorus (TP); (ii) microbiological variables, such as fecal coliforms (FC), total coliforms (TC), or *Escherichia coli* (*E. coli*) [8–10]. Relative recently, Aatik et al. developed a wastewater quality index (WWQI) to estimate the overall quality status of the raw and treated wastewaters. This is a number that can cumulatively describe the quality of an aggregate set of measured physicochemical and biological parameters [10]. For example the (waste)water can be described as: (i) poor for WWQI between 0 and 44; (ii) marginal for WWQI from 45 to 64; (iii) fair for WWQI from 65 to 79; (iv) good for WWQI from 80 to 94, and (v) excellent for WWQI from 95 to 100 [10,11].

Throughout all the stages in the slaughterhouse production line, large amounts of water are consumed (15–20 L/chicken), thus also requiring an intense processing of resulting wastewater. This is loaded with organic matter that may re-enter the environment in large amounts, via effluents [1,12]. It is therefore essential to apply efficient purification treatments of wastewater prior to release. Wastewater management is considered an important task, subjected to strict global regulations [13–15]. Some well-established physicochemical methods, used to treat the wastewater by removing contaminants such as metals, organic matter (OM) and total suspended solids, are [16]: coagulation–flocculation [17,18]; flotation [19,20]; electrocoagulation [21,22]; chemical precipitation [1,17,23–25]; ultra- and nano-membrane filtration and reverse osmosis [26,27]. Traditionally, biological (aerobic and anaerobic) treatment methods have been used for poultry slaughterhouse wastewater treatment. Unfortunately, both biological techniques present some limitations [28]. The conventional treatment processes of agro-industrial wastewater based on the aerobic technologies requires high energy consumption for aeration and generates large amounts of sludge. In recent years, these tend to have been replaced by more environmentally friendly anaerobic biotechnologies. Among others, for the slaughterhouse wastewater treatment, the sequencing batch reactor (SBR) showed a good efficiency for pollutant removal [15,28,29].

Classical measurements of wastewater properties based on determining the electrical conductivity, TSSs, TDSSs, turbidity, pH or COD values are far from an accurate description. Moreover, in many situations, within experimental errors, the samples cannot be differentiated using only these parameters. By comparison with usual situations, anyone can imagine that, knowing only the average age, it is difficult to describe the population structure of a country or, knowing the average wage, it is difficult to guess the distribution of income for the employees of a company. Similarly, knowing just a single value of a parameter, describing globally a complex sample with many components like wastewater, can dramatically limit any improvement in treatment methods. In these situations, improvement very often comes from a strategy of trial and error. Conversely, implementing characterization methods capable of discriminating between different components, and characterizing them, allows an educated strategy for an efficient improvement of treatment by knowing the mechanism of interaction between the treatment agent and sample (wastewater in this particular case). Such methods are based on Laplace spectroscopy, such as NMR relaxometry and diffusometry, or classical Fourier spectroscopy, such as FT-IR or UV-VIS.

NMR is one of the most widely used characterization methods for a large variety of materials, from plastics to elastomers [30–33], ordered tissues [34,35], denaturation of keratin [36,37], proton exchange membranes [38,39], biomaterials [40], construction materials [41] or even for in vivo human tissues [42]. With the exception of plastic/elastomers, for all other materials ^{1}H NMR implies the study of water states. H_2O is seen as a group of "spy" molecules in interaction with organic and inorganic materials. A logical step was to apply the well-known ^{1}H NMR methods, such as: (i) one-dimensional ^{1}H NMR relaxometry, to obtain the transverse relaxation times T_2-distribution, sensitive to water pools dimension [13,41,43,44]; (ii) one-dimensional ^{1}H NMR diffusometry to obtain the distribution of self-diffusion coefficient D, sensitive to molecular water mobility and water pools restrictions [31]. These two NMR parameters are sensitive to water impurities, and thus different water components can be revealed. For example, in slaughterhouse wastewater can be found: (i) soluble impurities that will increase the wastewater viscosity reducing the T_2 and D values or (ii) insoluble impurities that, depending on their dimensions, can be seen as different relaxation environments leading to multiple T_2-values. Having different sizes and different hydrophilicity (or hydrophobicity), such impurities can act (or not) as transport vehicles for water molecules. Then, one can observe multiple values in the measured D-distribution. The NMR data are analyzed by the inverse Laplace transform (ILT) [45–48]. In this way, one can obtain the distribution of relevant NMR parameters to describe the wastewater states by water components [13,41,43,44,49–52].

The aim of this study is to assess the efficiency of new methods, such as low field ^{1}H NMR relaxometry and diffusometry and VIS-nearIR spectroscopy in analyzing the polluted wastewater (PSW) during the treatment process in the slaughterhouse. The advantage of these methods resides in their ability to offer a large range (distribution) of the measured parameters. The sample apparent color is used to characterize the wastewater contamination with impurities. Such soluble or insoluble impurities can increase the turbidity of wastewater samples. In this study, we aim to better characterize these two parameters by VIS-nearIR spectra measured for a range of wavelengths from 390 to 980 nm. Afterwards, we show how the non-classical methods can be combined with the classical methods which characterize the entire wastewater samples, for a complex analysis using Principal Component (PCA) [53,54]. Thus, PCA statistical data lead to graphic plots that are easily interpreted to discriminate between all types of wastewater (at least between the polluted and less polluted). Thus, the PCA can be used to evaluate the treatment process efficiency and can offer information about the relevance of each measured parameter in the characterization of the PSW purification. Moreover, it can be used in combination with machine learning algorithms to provide predictions related to the stage of the treatment process. In this way, the integrated combination of NMR relaxometry and diffusiometry with VIS-nearIR spectroscopy and conventional measurements, such as pH, EC and TDSS in PCA analysis, enhanced by predictions performed by an ANN machine learning algorithm, will provide a comprehensive assessment of wastewater properties. This integrative approach ensures a thorough evaluation of the treatment process at various stages!

2. Materials and Methods

2.1. Materials and Processes in the Slaughterhouse Wastewater Purification Treatment

The wastewater and sludge samples were collected from a chicken slaughterhouse located in the northwestern part of Romania, from winter to spring in four different months. The main components of chicken slaughterhouse purification treatment plant are presented in Figure 1. The untreated wastewater was collected from the pumping basin where the wastewater is brought directly from the slaughterhouse. This wastewater contains all the specific organic matter pollutants resulting from the slaughterhouse, such as dejection, fodder, gastric contents, blood, fragments of meat and intestines, fats, feather remnants, etc. From here, the wastewater is pumped by a parabolic filter that allows only the wastewater and small particles to pass, while the large residues are collected by gravitation in an external tank (see Figure 1). Next, the wastewater is stored in a large pool, where it is periodically mixed by aeration to avoid the sedimentation of heavy matter. From here, the wastewater is pumped in a flotation basin for chemical treatment. This treatment is performed in the presence of SUPERFLOC C-2240 (registered trademark) liquid cationic polymer. According to the producers, this liquid coagulant polymer acts as primary coagulant, being a charge neutralization agent in the liquid/solid separation processes. The liquid polymer enables filtration, flotation of dissolved air, wastewater clarification, and reduces the use of inorganic salts. It is used at a ratio of approximately 1.8 L of liquid polymer to 1000 L of wastewater, but the effective dose is pH controlled.

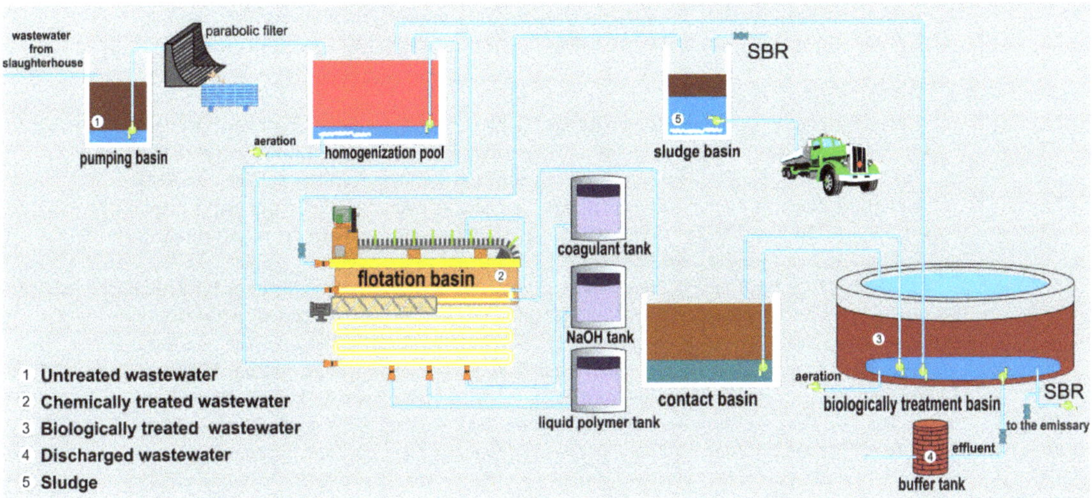

Figure 1. The main components of the wastewater purification treatment plant of a chicken slaughterhouse.

The second chemical agent is sodium hydroxide (NaOH), popularly known as caustic soda (or lye). This is a highly caustic base, used to decompose the proteins remained in the wastewater. Ferric sulphate is the third chemical agent, typically used for wastewater treatment and sludge conditioning, for removing phosphorus, to reduce the hydrogen sulphate, and to attenuate the corrosiveness and odor of wastewater and sludge. Being lighter than water, the coagulated sludge is collected from the wastewater surface in the flotation basin with a paddled conveyer belt and is stored in a sludge basin. From here, it is then transported to another city to be used in producing biogas.

The sludge samples were collected from the sludge basins, as shown in Figure 1. The chemically treated wastewater samples were collected from the bottom of the flotation basin through a drain hole (see Figure 1). After its separation from sludge, the chemically treated wastewater is pumped in the contact tank that communicates with an external basin where the liquid undergoes the biological treatment process. This is mainly based on the anaerobic bacterial treatment in a classical sequencing batch reactor, also shown in Figure 1. In this biological treatment basin, the bacterial population is controlled by wastewater retraction in the contact basin and/or by aeration. The samples of biologically treated wastewater were collected in daylight during the purification process. The bio-purification technology implies numerous steps. To be discharged after a sequence of nitrification, the wastewater is subjected to a 2 h stage of nitro-sedimentation, followed by another 2 h stage of sedimentation. The control mechanism of the wastewater station only allows the wastewater to be discharged after these 4 h of sedimentation stage, which usually takes place overnight. The samples of discharged, and hence completely treated wastewater, were collected from a buffer tank located between the slaughterhouse treatment plant and the Crasna River flowing nearby. The sediments collected at the bottom of the biological treatment basin are pumped back in the homogenization pool, and re-enter to the purification circuit, again undergoing chemical treatment and separation from sludge in the flotation basin. Thus, the sludge results are combined from both the biological and the chemical treatments.

The wastewater samples (untreated wastewater, chemically treated wastewater, biologically treated wastewater, discharged wastewater) and sludge were each stored each in ½ L plastic containers, immediately transported to the analytical laboratory at the Technical University of Cluj-Napoca (TUCN), and the analyses run as soon as possible (1 to 3 days). Prior to each measurement, the samples were homogenized by manually agitating the bottles for 1–3 min. Between the successive sets of measurements the wastewater samples were stored in the transport bottles at room temperature and protected from light.

2.2. Methods

The first set of measurements were performed for the overall parameters: pH, electrical conductivity and totally dissolved solids. The second set of measurements consisted of the ^{1}H NMR relaxometry that lasted several minutes per measurement. The ^{1}H NMR diffusometry measurements were the longest, lasting approximately 80 min per sample. Finally, the FT-IR spectra (not discussed here) were the last performed measurements, requiring in total 5 min per sample [13].

The ^{1}H NMR relaxometry and diffusometry measurements for all the wastewater samples were performed using a Bruker Minispec MQ 20 (Bruker Co., Ettlingen, Germany) spectrometer working at 19.69 MHz. The NMR relaxation data were recorded using the well-known CPMG (Carr–Purcell–Meiboom–Gill) pulse sequence [13,41,42] with echo time $2\tau = 2$ ms (Figure S1a, in Supplementary Information). A total of 3000 echoes were registered, and the repetition time or recycle delay (RD) was set at 3 s. For a good signal-to-noise ratio (Figure 2a), acquiring 32 scans was sufficient [13]. Then, the experimental data were processed using a fast inverse Laplace-like transform (ILT) algorithm, increasingly used in recent years to analyze multi-exponential decay curves [13,42,44–46]. The NMR self-diffusion data were recorded using the pulsed gradient stimulated echo (PGSE) pulse sequence (Figure S1b in Supplementary Information). The inter echo time τ was set at 3 ms, the encoding and decoding pulse gradients duration δ was 0.4 ms, while the self-diffusion time Δ was set at 20 ms. A number of 64 acquisitions were collected with a recycle delay of 1.5 s. The recorded data were analyzed by the ILT algorithm with the proper (see below) kernel. The ILT analysis [45–48] will be described in Section 2.3 below.

The Vis-nearIR spectra were recorded in less than 10 s per sample, using a PASCO spectrometer. For some wastewater samples, a sedimentation time was necessary before each measurement. The sludge samples were distilled at a ratio of 1:20, by adding 1 mL sludge to 20 mL of distilled water. The measurement set-up was as follows: (i) the lower wavelength λ_L was 379.928 nm; (ii) the higher wavelength λ_H was 950.11 nm; (iii) a total number of 2132 points were acquired with a resolution of 0.268 nm; (iv) 25 scans were accumulated to increase the signal-to-noise ratio (SNR).

The pH and the electrical conductivity were measured using a water quality meter multi-parameter Model 8603. The totally dissolved solids were measured using a pocket TDSS/EC meter. All the methods of measurement/analysis follow the standard protocols [1].

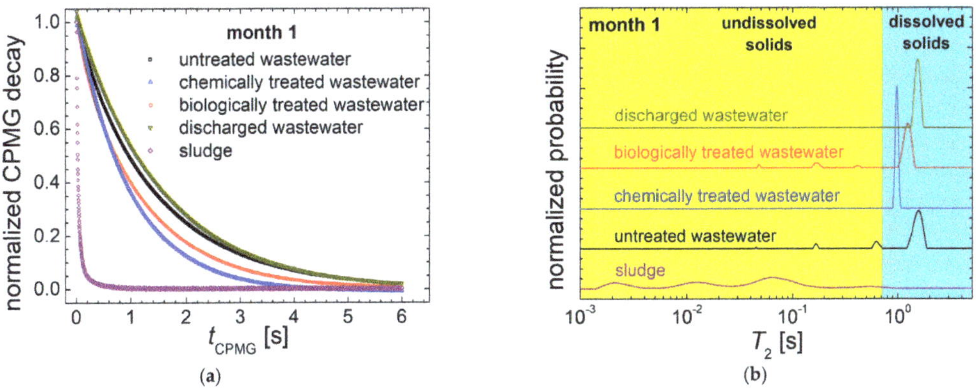

Figure 2. (**a**) The normalized CPMG decays recorded for the slaughterhouse wastewater (untreated–black square, chemically treated–blue upper triangle, biologically treated–red circle and discharged wastewater–dark yellow lower triangle) and sludge (diamonds) collected in month 1 of monitoring; (**b**) the normalized T_2-distributions of the CPMG decay curves presented in (**a**).

2.3. Data Analysis

The CPMG decays of nuclear magnetization, $M(t_{\mathrm{CPMG}})$ characteristic of water with many types of impurities (see Figure 2a), are assumed to be multi-exponential. Each specific transverse relaxation time describes a certain spin dynamics according to the ^{1}H environment. Therefore, the $M(t_{\mathrm{CPMG}})$ decay can be written as [42–44]:

$$M(t_{\mathrm{CPMG}}) = \sum_{i=1}^{N} P(T_{2,\mathrm{i}})exp\left\{-\frac{t_{\mathrm{CPMG}}}{T_{2,\mathrm{i}}}\right\}, \tag{1}$$

where $P(T_{2,\mathrm{i}})$ is the probability of finding a statistical spin ensemble (water ^{1}H) characterized by a specific transverse relaxation time $T_{2,\mathrm{i}}$ [36]. $t_{\mathrm{CPMG}} = \tau_2 = 2n\tau$ is the total duration of the CPMG experiment, as can be seen from Figure S1a in the Supplementary Information. To consider a real distribution of transverse relaxation times, the decay of the nuclear magnetization presented in Equation (1) can be considered to have an integral Laplace-like form that also contains the distribution function of transverse relaxation times $f(T_2)$ [42–48]:

$$F(t) = \int_0^{+\infty} f(T_2)exp\left\{-\frac{t}{T_2}\right\}dT_2. \tag{2}$$

Nevertheless, the practical way to analyze the discretized acquired NMR signal is to combine the finite summation presented in Equation (1) and the distribution function $f(T_2)$. Finally, the goal of an inverse Laplace analysis is to find a normalized distribution function $f(T_2)$. This will be interpreted as normalized probability to find a statistically relevant number of water ^{1}H characterized by a particular value of T_2. Here, we have to remark that the transverse relaxation process assumes a certain loss of phase of the nuclear spins processing around a static magnetic field, and at the same time being in interaction with neighboring spins. Therefore, for just an isolated spin, the T_2 is not defined.

Similarly, the decay of the normalized pulsed field gradient echo (PGSE) NMR signal can be viewed in an integral Laplace-like form containing the specific distribution function $f(D)$ of the self-diffusion coefficient [41,46]:

$$\frac{\mathrm{S}(g)}{\mathrm{S}(0)} = \int_0^{+\infty} f(D)\mathrm{exp}\left\{-\gamma^2 g^2 \delta^2 D\left(\Delta - \frac{\delta}{3}\right)\right\}\mathrm{d}D = \int_0^{+\infty} f(D)\mathrm{exp}\{-\mathrm{b}D\}dD, \tag{3}$$

where D is the self-diffusion coefficient and the distribution function $f(D)$ is the parameter function to be obtained. The new kernel of Laplace-like integral presented in Equation (3) contains the ^1H gyromagnetic ratio γ; the magnetic field gradient strength g; the duration of encoding/decoding pulsed gradients δ and the self-diffusion time Δ.

2.4. Statistical Analysis

In general it is not possible to directly compare the behavioral effects of two parameters of different types, though a standard method is largely used to analyze the statistical data (usually from more than one group) that are characterized by numerous variables or observables having different measurement units. This is multivariate analysis, in particular, principal components analysis [54]. For our study, we implemented our own numerical program in MatLab 2020 and we plotted the results in Microsoft Excel software 2024. For such PCA statistical analysis, a data matrix is produced containing the input values that will be discussed later in Table 1. As variables, we selected: (i) the total absorbance from the Vis-nearIR spectra; (ii) the most probable transverse relaxation time $T_{2,1}$ characteristic to wastewater with dissolved solids, from ^1H NMR relaxometry measurements; (iii) the most probable self-diffusion coefficient D_1 specific also to wastewater with dissolved solids, from ^1H NMR diffusometry measurements; and the bulk values for (iv) pH; (v) electrical conductivity and (vi) totally dissolved solids. All the values of these six parameters were considered for all our five groups of four samples, each of them corresponding to one month of monitoring. The values of these parameters form the input matrix with 6 columns and 20 rows. The results of the PCA method will be largely discussed later in the text.

Table 1. The values of relevant parameters (total VIS-nearIR absorbance, the most probable transverse relaxation time $T_{2,1}$, the self-diffusion coefficient D_1, pH, electrical conductivity EC and the total dissolved solids TDSS) measured for the slaughterhouse wastewater (untreated, chemically treated, biologically treated and discharged) and sludge collected during four months of monitoring.

Wastewater	Month	Total VIS-NearIR Absorbance	$T_{2,1}$ [s]	D_1 [10^{-9} m^2/s]	pH [-]	EC [µS/cm]	TDSS [ppm]
untreated wastewater	1	1.06 *	1.57	2.76	5.11	478	263
	2	0.14	1.71	2.86	6.75	1596	795
	3	1.40 *	1.31	2.78	7.16	1236	618
	4	2.13 *	1.59	2.81	7.17	658	658
chemically treated wastewater	1	0.06	0.96	3.03	8.4	947	470
	2	0.04	1.36	2.88	6.03	860	484
	3	0.05	1.02	2.90	6.35	740	374
	4	0.02	0.89	2.95	6.53	383	383
biologically treated wastewater	1	0.33 **	1.24	2.68	7.15	1182	597
	2	0.14 **	0.58	2.81	7.29	822	412
	3	2.99 **	0.47	2.68	6.25	686	344
	4	0.04 **	0.33	2.67	6.66	353	353
discharged wastewater	1	0.02	1.53	2.98	7.52	860	433
	2	0.02	1.30	2.86	5.84	429	429
	3	0.03	0.99	2.91	6.56	704	352
	4	0.002	1.04	2.88	6.61	356	356
Sludge	1	1.39 ***	0.52	2.31	7.23	2880	1662
	2	0.33 ***	0.32	2.62	6.92	2070	1045
	3	0.43 ***	0.34	2.66	6.13	1668	826
	4	0.32 ***	0.14	2.60	6.94	963	963

Notes: * Measured after 30 min of sedimentation; ** measured after 10 min of sedimentation; *** measured after 5 min of sedimentation; in all cases the measurement errors are smaller than 5%.

2.5. Prediction Using Machine Learning ML5, an AI-Based Algorithm

The two-dimensional PC1-PC2 map obtained from PCA analysis were used for an advanced prediction of probability maps from any newly assumed measurements using the machine learning library ML5 [55,56]. For this, a dedicated program was written in JavaScript following the descriptions presented in [56] and adapted for our purpose. The 20 (=5 × 4) values obtained from PCA analysis was used to train an Artificial Neural Network (ANN) with a learning rate of 10% and for 10,000 epochs. The resulting model was saved and reloaded to predict the probability of a new measurement, located on each element of a 100 × 100 matrix correlated to the PC1-PC2 map obtained from the PCA analysis, associated with each type of wastewater or sludge.

3. Results

3.1. ^{1}H NMR Relaxometry of Wastewater and of Sludge from the Chicken Slaughterhouse

The results obtained by measuring the normalized ^{1}H nuclear magnetization decays during the CPMG pulse sequence for all types of wastewater and sludge samples collected over the first month of monitoring are compared in Figure 2a. At 6 s, a complete decay (less than 5% from initial magnetization) is observed for all the wastewater samples, indicating a large mobility of water molecules. An equivalent decay is observed for sludge (purple diamond in Figure 2a), but this is achieved in less than 0.2 s, indicating, as expected, a strong mobility limitation for the water molecules. Among wastewater samples, the slowest decay was measured in discharged wastewater (dark yellow down triangle), which is expected to present the smallest number of impurities.

Chemically (blue upper triangle) and biologically (red circle) treated wastewaters present the fastest decays, suggesting the presence of large amounts of undissolved solids and/or dissolved solids, and/or the presence of paramagnetic impurities.

A much better analysis of the normalized ^{1}H nuclear magnetization decays, providing deeper information about water state and efficiency of treatment process, can be achieved by an inverse Laplace analysis [42–48]. The normalized transverse relaxation time T_2-distributions obtained from ILT of CPMG decays measured for slaughterhouse wastewater (untreated, chemically and biologically treated and discharged wastewater) and for sludge are presented in Figure 2b. The normalized probability function $f(T_2)$ presents a series of peaks at different T_2–values covering more than three orders of magnitude from 1 ms up to 3 s. From left to right (from small to large T_2–values) these peaks can be associated with water molecules being in different environments. For example, the ^{1}H with more restricted mobility are characterized by small T_2–values, while water ^{1}H with less restricted mobility are characterized by large T_2–values. Considering the specificity of our samples, one can divide the full domain of T_2–values in two main subdomains. Thus, there is the range of T_2–values from 1 ms up to 700 ms, of water molecules attached to undissolved solids (on yellow background), and the subdomain of T_2–values from 700 ms to 3 s (with blue-cyan background) where the particular position of peaks and therefore of the most probable T_2–values is influenced by the type and number of dissolved particles [13]. In particular, we measured for our low filed NMR spectrometer a T_2–value for distilled water (no dissolved or undissolved particles) of 2.66 ± 0.1 s (see Figure 3). The presence of various dissolved solids should be then evaluated by taking into account this reference value. From Figure 2b, for wastewater samples in the subdomain of water pools with dissolved particles, only one peak can be observed, but up to three peaks for water pools with undissolved particles. It is reasonable to assume that the water molecules are attached permanently or can experience a certain exchange between free water and bound water [41] and the attachment is to undissolved particles of different dimensions. The size (and shape, density, and hydrophilicity/hydrophobicity) of such particles can influence the mobility of undissolved particles, hence the mobility of water molecules attached to these particles. The NMR T_2 relaxation time is sensitive to ^{1}H mobility and therefore it is a good parameter to sense and characterize the presence and sizes of such undissolved particles. Furthermore, it is reasonable to assume the fact that large particles will have a small mobility and the corresponding T_2 relaxation times values are also small. Small particles will have then a large mobility and a large corresponding T_2–value. The integral area under the peaks, as it is measured from the T_2-distributions, is proportional with the number of ^1H present in each pool with a specific transverse relaxation time value.

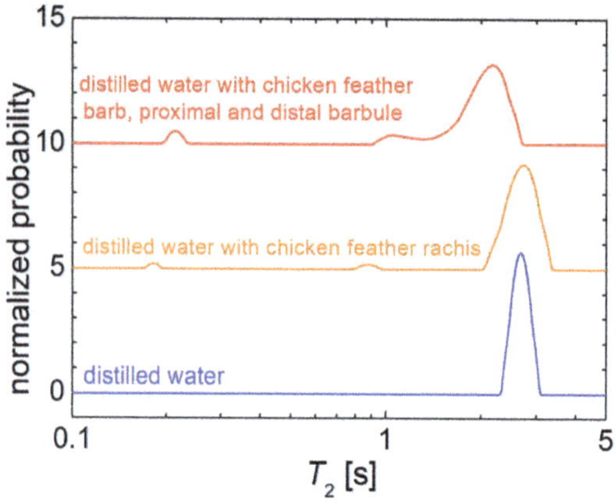

Figure 3. The normalized T_2-distributions recorded for distilled water (bottom blue line), distilled water with chickens feather rachis (middle orange line) and distilled water with chicken feather barb and proximal and distal barbule (top red line).

Considering the above, in Figure 2b it can be observed that the T_2-distributions measured for the untreated wastewater is characterized by four peaks: (i) a large peak (the largest integral area under the peak) in the subdomain of water with dissolved solids with the most probable T_2–value of 1.57 s not so far from the limit of 2.66 s (blue curve in Figure 3), indicating a small number of dissolved solids; (ii) a relatively large peak centered at a T_2–value of ~0.63 s indicating a large number of small undissolved particles; (iii) a relatively medium size peak centered at a T_2–value of ~0.17 s indicating the presence of medium size undissolved particles and (iv) a small (at the observable limit) peak centered at a T_2–value of ~45 ms indicating the presence of relatively large undissolved particles. In general, ^1H NMR relaxometry is also a method capable, within certain limits, of offering the size of such particles (or pores [36]), but the experiments need to be performed under more idealistic conditions for inorganic interface water-matter, where a certain surface relaxation parameter has to be estimated, conditions that are not fulfilled in our study. Furthermore, for untreated wastewater, the sampling procedure can play an essential role. For example, in our case, the narrow bore pipette used to insert the sample in the NMR tube was acting as a filter for the larger undissolved fragments that are typically present in untreated wastewater.

The chemically treated wastewater (blue line in Figure 2b) is characterized by a single peak in the subdomains of water with dissolved solids. The fact that no peaks were found in the domain of water with undissolved solids is a good indicator of the efficiency of the coagulation–flotation sludge separation process applied in the chemical treatment basin. On the other hand, the most probable T_2–value of ~0.96 s indicates a larger presence of dissolved solids compared with the T_2–value of ~1.57 s measured for the untreated wastewater. Two processes could explain these findings: (i) the undissolved solids subjected to chemical treatment became soluble in water and (ii) the chemicals (liquid polymer, caustic soda and especially the ferric sulphate) enrich the water with paramagnetic impurities. The measured signal is the narrowest peak, compared to all the peaks measured for wastewater samples, also indicating a high degree of homogeneity of this sample. Surprisingly, the biologically treated wastewater is characterized again by four peaks, one in the subdomain of water with dissolved solids and three in the range of water with undissolved solids. At first glance this looks like a regression in the treatment process, but one must consider the complex mechanism of wastewater treatment presented in Figure 1. This wastewater sample was collected from the biological treatment basin, in daylight and in the presence of anaerobic bacteria, with nitrification and aeration processes that mixed all components, prior to the nitro-sedimentation and sedimentation. The peaks position in the T_2-distributions measured for the biologically treated wastewater (red distribution function) for water with undissolved solids is almost identical to the position of the peaks obtained for untreated wastewater (black curve). A quality improvement in the water with dissolved solids can be noticed by observing that the most probable T_2–value increases from ~0.96 s (measured for chemically treated wastewater) to ~1.24 s.

The efficiency of the pollutants removal using this sequencing batch reactor for wastewater treatment is observed from the T_2-distributions measured for the discharged wastewater, after several hours allowing for nitro-sedimentation and sedimentation. The T_2-distribution is characterized by a single peak located in the subdomain of water with dissolved solids at a T_2–value of ~1.53 s. The position of this peak is similar to the position of the peak obtained for untreated wastewater sample in the subdomain of water with dissolved solids. Its narrower linewidth indicates less variation in the types of dissolved solids.

The sludge (purple distribution at the bottom of Figure 2b) also presents four peaks, but all of them are characterized by reduced T_2–values, suggesting, as expected, the presence of protons with a reduced mobility. In fact, the term 'water with undissolved solids'; may by inappropriate in this case, since the sludge is a complex colloidal system [44]. Here, a large amount of water is trapped in the organic and inorganic matter, and therefore has a reduced mobility and cannot be easily released. Moreover, taking into consideration the lower limit of T_2–values (~2 ms) measured for this sample of slaughterhouse sludge, one can reasonably assume that this peak corresponds to the ^{1}H from organic matter. The remaining peaks in the range of ~6 ms up to 1 s can be associated with water strongly bound to the organic phase via vicinal water (hydrogen bound), with interstitial water physically trapped in bio-floc, by steric hindrance and with a small amount of water with higher mobility being less affected by the solid composition [43,44].

To assess the effect of specific solids on the T_2-distributions measured for this slaughterhouse wastewater, three samples were prepared: (i) distilled water; (ii) distilled water with chicken feather rachis and (iii) distilled water with chicken feather barb, proximal and distal barbule. The chicken feather was used as it was collected without additional purification by washing and/or sterilization. The measured T_2-distributions for these three samples are presented in Figure 3. It can be observed that the sample of distilled water (blue distribution) presents a single peak with the most probable T_2–value (maximum of peak) centered at 2.66 s. The T_2-distributions measured for distilled water samples with parts of chicken feather present three peaks: (i) the main broad peak (compared to the peak of pure distilled water) indicates an increased inhomogeneity of the aqueous medium; (ii) a small peak located between ~0.8 and 1.3 s placed in the subdomain of water with dissolved solids indicates that part of the feather impurities are dissolved in water and (iii) a small peak located between ~0.16 and ~0.23 s, indicates a less restricted mobility. This can be associated with water attached to chicken feather parts. This water may also experience a certain exchange with free water, therefore the real T_2–values may be affected by this exchange process [47]. As expected, due to smaller parts of feather, such as barb, proximal and distal barbule compared to rachis, the T_2-distributions measured for the first sample (top, red distribution in Figure 3) present broader peaks indicating a higher degree of inhomogeneity. Moreover, the position of the main peak is shifted towards smaller values indicating a reduction in the water mobility, most probably, by the interstitial water between barbs and barbules. In conclusion, these measurements demonstrates the effect of typical pollutants present in the chicken slaughterhouse wastewater, able to induce additional peaks at smaller T_2–values in the T_2–subdomains of water with undissolved solids (see Figure 2b), but also to shift the peaks located in the T_2–subdomains of water with dissolved solids towards smaller T_2–values.

The normalized T_2-distributions recorded for four types of slaughterhouse wastewater (untreated, chemically treated, biologically treated and discharged water, respectively) collected during months 2, 3 and 4 are presented in Figure 4. As expected, the main characteristics of these T_2-distributions are similar to those presented by samples collected in month 1, shown in Figure 2b, and discussed above. Nevertheless, the existence of some variability indicates the necessity for a statistical analysis of such samples. In this sense, for the T_2-distributions collected during months 2 to 4, one can remark: (i) the presence of two to three peaks for the untreated wastewater; (ii) the presence of one or no-peak for the chemically treated wastewater; (iii) two or three peaks for the biologically treated wastewater and (iv) just one peak for the discharged wastewater, which confirms the efficiency of the wastewater treatment applied.

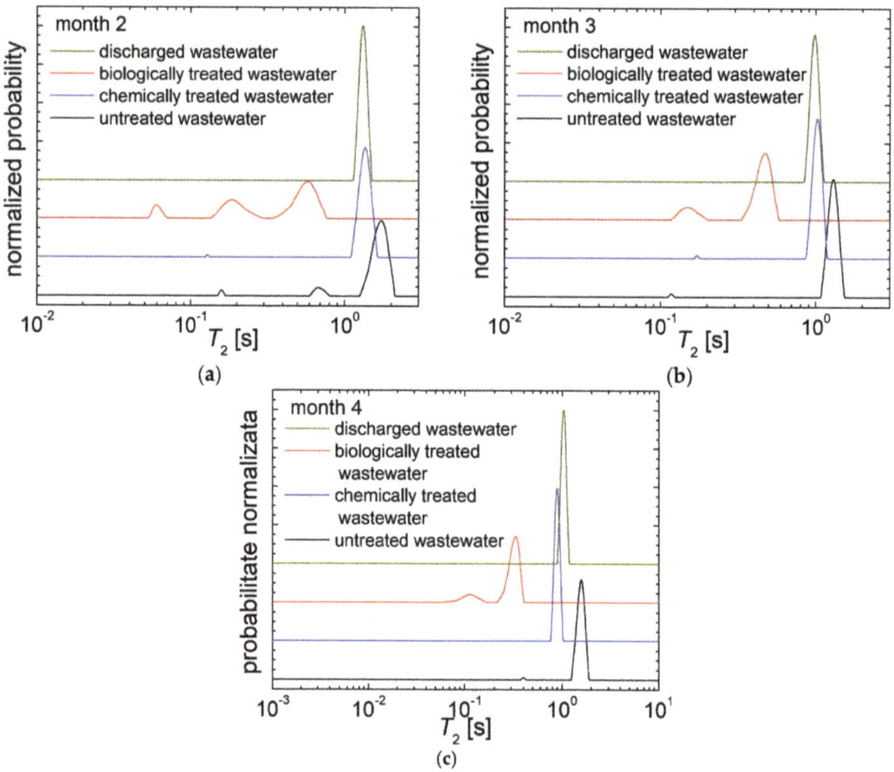

Figure 4. The normalized T_2-distributions recorded for the slaughterhouse wastewater: untreated (black), chemically treated (blue), biologically treated (red) and discharged wastewater (dark yellow) collected in (**a**) month 2; (**b**) month 3; (**c**) month 4.

Since the discharged wastewater presents only one peak, identified as the main peak located at the largest T_2-values, only the most probable values measured for this peak (here forth named $T_{2,1}$), associated with water containing dissolved solids, are comparatively presented in Figure 5a. This is the case for all four types of wastewater and sludge collected from the chicken slaughterhouse in all four months of monitoring. Relatively large variations are observed for this NMR parameter, especially in samples of biologically treated wastewater and sludge. The normalized T_2-distributions of the last one are presented in Figure S2 from Supplementary Information. The average values and the standard deviations calculated for those four months of observations are presented in Figure 5b). The largest value of average $T_{2,1}$ was calculated for the untreated wastewater (~1.54 s), indicating a small number of dissolved solids in this samples. The relatively small standard deviation values indicate a certain consistency and homogeneity of these samples. The average value of $T_{2,1}$ calculated for the discharged wastewater (~1.22 s) has a relatively large standard deviation compared to the untreated wastewater samples. The larger variability presented by these samples can indicate the degree of the wastewater treatment efficiency. The average $T_{2,1}$ value for the chemically treated wastewater (~1.06 s) is calculated with a comparable relative measurement error. However, considering the large overlaps of error bar domains observed for discharged and chemically treated wastewater, one can say that this parameter cannot differentiate between these two types of wastewater. Among the wastewater samples, the biologically treated wastewater showed the smallest $T_{2,1}$ value (~0.66 s) and the largest standard deviations. This clearly indicates the presence of large amounts of dissolved solid in wastewater during the biological treatment stage. As expected, the smallest averaged $T_{2,1}$ value was calculated for the sludge (~0.33 s). For these samples the relatively large standard deviation values indicate their large variety. Since the untreated wastewater has an average $T_{2,1}$ value larger than that calculated for discharged wastewater, and due to the fact that the average value of $T_{2,1}$ calculated for chemically treated wastewater is, within the experimental error limit, comparable with that for the discharged wastewater, this parameter ($T_{2,1}$) alone is not a good indicator of water quality. Therefore, other parameters have to be considered, such as the ^{1}H NMR self-diffusion coefficient.

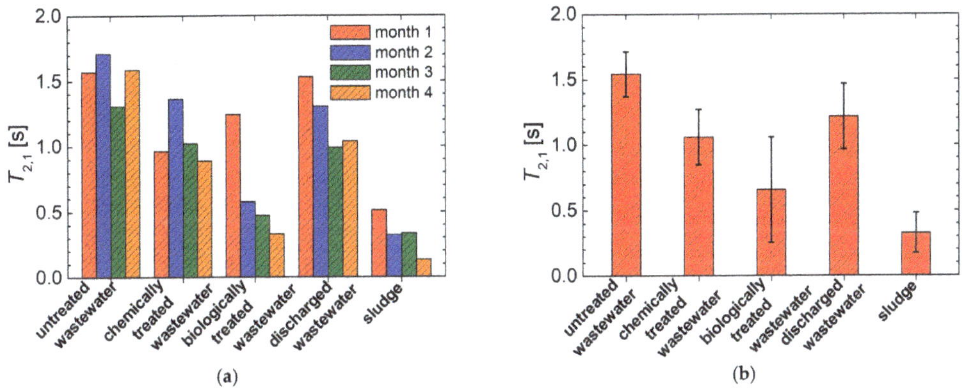

Figure 5. (**a**) The most probable T_2 values measured from the T_2-distribution as maximum of the main peak for all wastewater samples (untreated, chemically and biologically treated and discharged wastewater) and sludge collected from a slaughterhouse in months 1 to 4; (**b**) the average over all months of measurement of the most probable T_2 values presented in (**a**).

3.2. ^{1}H NMR Diffusometry of Wastewater and Sludge from Chicken Slaughterhouse

The ^{1}H NMR self-diffusion coefficient is a sensitive NMR parameter that can describe well the properties of fluids located in pores, capillaries, or in every case in which dissolved particles can influence the fluid molecules' mobility, as in our case. Figure 6a shows (on a logarithmic horizontal scale) the PGSE decays measured for all four types of slaughterhouse wastewater and sludge. As expected, the decay measured for sludge is characterized by the smallest diffusion (due to the strongest limitation of water molecules' mobility given by their interaction with large amounts of pollutants) observed as the slowest decaying curve. Such curves are hard to be interpreted quantitatively; therefore, a Laplace-like analysis is performed using the kernel presented in Equation (3). The normalized D-distribution functions resulting from this analysis are presented in Figure 6b. The domain of D-distributions had to be extended, compared to the normal domain for the self-diffusion coefficient [47]. This extension becomes reasonable by considering, as in the case of relaxation data (see Figure 2b), the split of D-distributions in two: (i) the first corresponding to the self-diffusion of bulk water in the samples (see the blue background in Figure 6b) ranging for our samples from 10^{-9} m^2/s up to (an arbitrary) 10^{-8} m^2/s; (ii) the second associated to water molecules attached to solid residues and transported by flotation and sedimentation, having therefore a larger apparent diffusion coefficient in the range of 10^{-8} m^2/s up to 10^{-6} m^2/s. In this case, the limit of separation between the real self-diffusion coefficient and the apparent diffusion coefficient, describing a flow rather than a self-diffusion process, was chosen as a convenient route between these two domains.

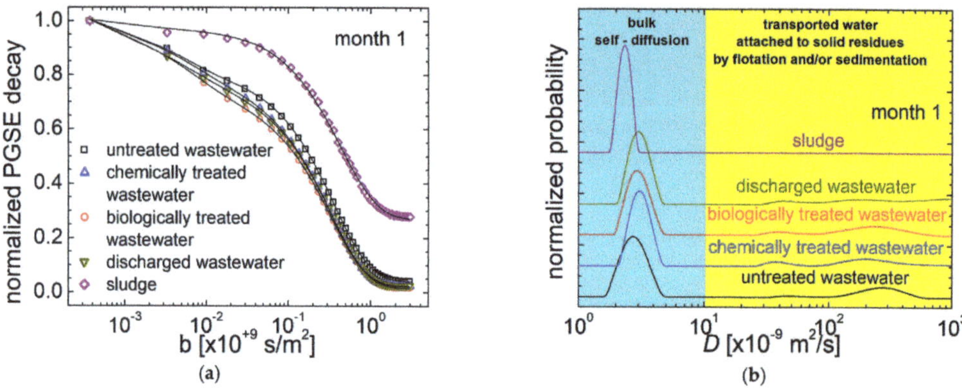

Figure 6. (**a**) The normalized PGSE decays recorded for the slaughterhouse wastewater (untreated—black square, chemically treated—blue upper triangle, biologically treated—red circle and discharged wastewater—dark yellow lower triangle) and sludge collected in month 1; (**b**) the normalized *D*-distributions of the PGSE decay curves presented in (**a**).

All samples present the largest amount of water as bulk water characterized by the largest peak in the self-diffusion coefficient distribution (Figure 6b). For the first month of monitoring, the wastewater samples (untreated, chemically and biologically treated and discharged wastewaters) presented three peaks, the largest, as mentioned, in the domain of self-diffusion and two of them in the domain of transported water. It is natural to assume that the largest transport coefficient should be associated with water characterized by a large velocity, and therefore a large mobility, specific to the small particles that the water is attached to. On the contrary, if the water molecules are attached to large particles, the mobility of this water-particle system is small. Therefore, the mobility of attached water is small and it is observed as a small apparent diffusion coefficient. The untreated wastewater leads to a broad peak centered at 2.76×10^{-9} m^2/s. This becomes narrow as a result of the purification process that reduces the samples' heterogeneity. A narrow peak was also measured for the sludge sample. The most probable self-diffusion coefficient (the peak maximum) increases from 2.76×10^{-9} m^2/s measured for untreated samples of wastewater to 2.76×10^{-9} m^2/s to 2.98×10^{-9} m^2/s for discharged wastewater, indicating an increased mobility (or reduced restrictions). A large value is obtained for chemically treated wastewater (3.03×10^{-9} m^2/s). The lowest most probable self-diffusion coefficient was measured for sludge (2.31×10^{-9} m^2/s). This is normal, since this sample contains the largest amount of residue. For the first month of monitoring, the *D*-distributions for sludge present no peaks in the domain of transported water. This is not due to the fact that there are no residues but, on the contrary, may be due to the fact that the amount of residue is so large that the mobility of water molecules, due to the sedimentation or flotation processes, is strongly reduced. For months of monitoring 2–4, the *D*-distributions measured for sludge samples presented similar peaks in the water transport domain (see Figure S3 from the Supplementary Information).

Two broad peaks with small amplitude are observed for all the wastewater samples in the transported water domain (Figure 6b—yellow background). In all cases, the integral area under the peaks corresponding to water molecules attached to small particles (the larger D-values, often of the order of hundreds of 10^{-9} m^2/s) is larger than the integral area of peaks corresponding to water molecules attached to large particles (the small D-values, often of the order of tens of 10^{-9} m^2/s). This indicates the presence of larger numbers of small particles in the slaughterhouse wastewater. The largest number of such small particles was found in untreated wastewater, as expected. The measurement of self-diffusion coefficient distributions for the discharged wastewater (Figure 6b) revealed peaks in the domain of transported water. This was a surprise, given that, for the relaxation times' T_2-distributions, the discharged wastewater generated a single peak (see Figures 2b and 4). The presence of several peaks in the D-distributions measured for discharged wastewater associated with water attached to mobile particles indicates that the ^{1}H NMR measurement of the self-diffusion coefficient is a more sensitive method than ^{1}H NMR relaxometry. The area under these peaks indicates the presence of a small number of such particles in the discharged wastewater. This probably has the same (or appropriate) relaxation time T_2 as the bulk water, therefore cannot be distinguished by the ^{1}H NMR relaxometry method. The peaks measured in the transported water domain are broad, indicating a large heterogeneity of samples, here translated into a large distribution of particle dimensions. Moreover, the presence of two resolved peaks (instead of just one really broad peak) suggests the presence of at least two types of particles with different origins.

The normalized self-diffusion D-distributions measured for all chicken slaughterhouse wastewater samples collected in months 2, 3 and 4 of monitoring are presented in Figure 7. Generally, the characteristics previously described for month 1 of monitoring are also valid for these samples. One can remark: (i) the similarly small deviation of the most probable self-diffusion coefficient (measured at maximum of peaks from D-distributions) that from here will be labeled as D_1; (ii) the large amount of water attached to small particles especially for untreated wastewater, chemically treated wastewater and discharged wastewater; (iii) a comparable or small amount of water attached to small particles (D of the order of hundreds of 10^{-9} m^2/s—from now labeled as D_3) than those attached to large particles (D of the order of tens of 10^{-9} m^2/s—from now labeled as D_2).

The measurement parameters (magnetic field gradients and delays in the PGSE pulse sequence—see Figure S1b from the Supplementary Information) are calibrated for good measurement of the bulk water self-diffusion coefficient. For our samples, this parameter (D_1) is measured with the largest precision (affecting the full decay curve—see Figure 6a) compared with the apparent diffusion coefficients D_2 and D_3. The latter two are measured with a small precision since, being one or two orders of measurement larger (than D_1), can only affect fewer points at the beginning of the decay PGSE curve.

For a quantitative analysis of the purification process via self-diffusion, only the D_1 diffusion coefficient will be considered. All D_1 values measured for all four types of wastewater samples and sludge for all four months of monitoring are comparatively presented in Figure 8a. With some exceptions (month 2 for untreated wastewater and biologically treated wastewater, and month 1 for chemically treated wastewater, discharged wastewater and sludge), the measured values of the most probable self-diffusion coefficient D_1 are similar for the other three months. The average of all four months and the measurement errors are presented in Figure 8b. Within the experimental error limit, one can observe that the D_1 parameter can be used to differentiate between the types of wastewater. An overlap only exists between the chemically treated wastewater and the discharged wastewater. The smallest D_1 value (and the largest error) was measured for sludge, and the largest were measured for chemically treated wastewater and for discharged wastewater. In conclusion, the large value measured of the self-diffusion coefficient D_1 for discharged wastewater indicates a small restriction of the water molecules' mobility and therefore a good efficiency of the wastewater purification applied in the chicken slaughterhouse.

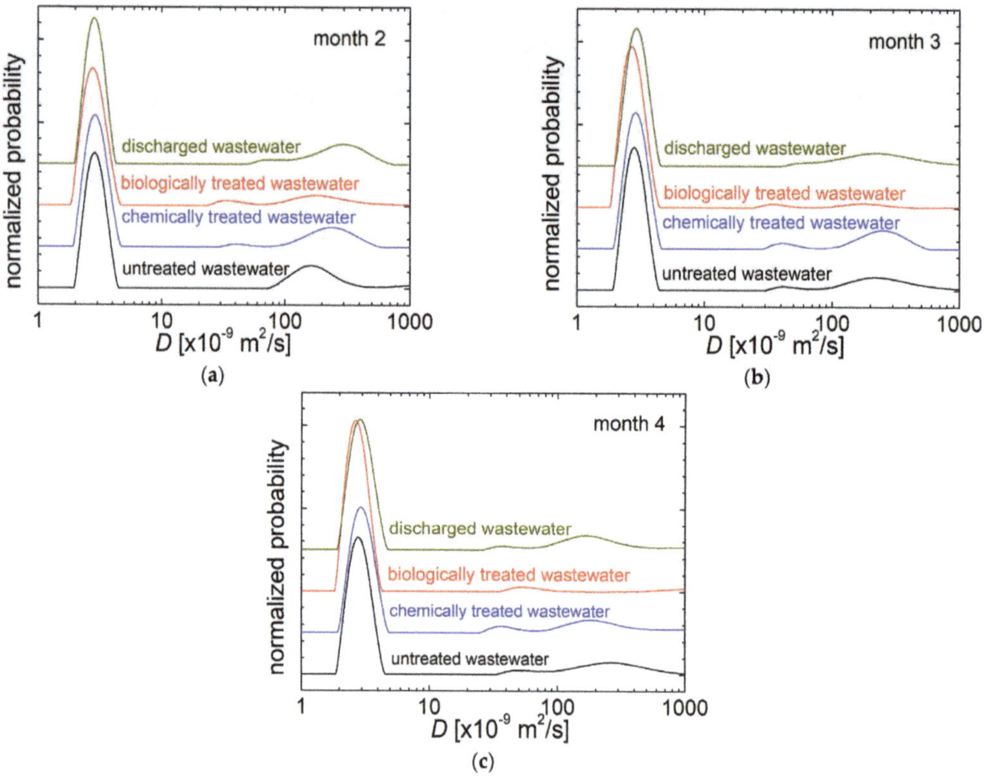

Figure 7. The normalized D–distributions recorded for the slaughterhouse wastewater: untreated (black), chemically treated (blue), biologically treated (red) and discharged wastewater (dark yellow) collected in (**a**) month 2; (**b**) month 3; (**c**) month 4 of monitoring.

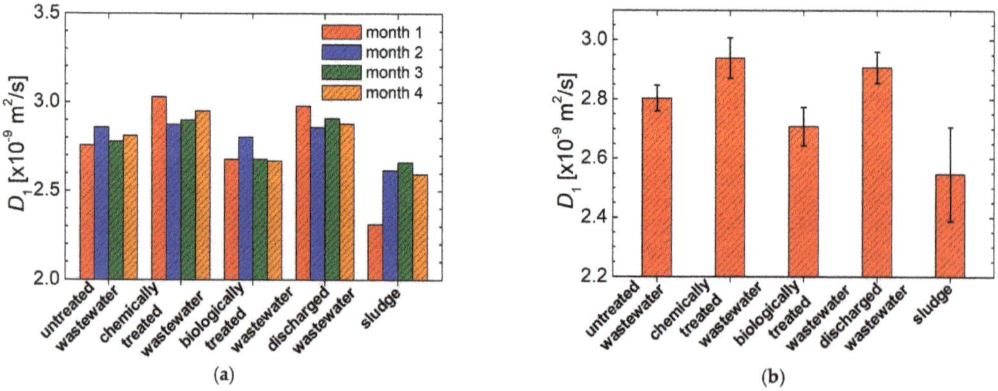

Figure 8. (**a**) The most probable self-diffusion coefficient D_1 values measured from the D–distribution as maximum of the main peak for all wastewater samples (untreated, chemically and biologically treated and discharged wastewater) and sludge collected from a slaughterhouse in months 1 to 4; (**b**) the average over all months of measurement of the most probable self-diffusion coefficient D_1 values presented in (**a**).

3.3. VIS-NearIR Spectroscopy

Both ^{1}H NMR relaxometry and diffusometry, as previously discussed, show that even in the discharged wastewater from chicken slaughterhouse, there is a variety of particles visible or invisible to the naked eye. These particles will affect (by absorption) the visible light that eventually passes through wastewater. The classical method largely used to quantify the amount of such particles is by determining its turbidity [3]. This is a global parameter, giving a single value for a specific sample. We propose to replace the turbidity measurement by acquiring a VIS-nearIR spectrum.

An example of such VIS-nearIR spectra recorded for untreated wastewater (after 30 min of sedimentation), chemically treated wastewater, biologically treated wastewater (after 10 min of sedimentation), discharged wastewater and sludge (1 mL of sludge distilled in 20 mL of distilled water and then measured after 5 min of sedimentation) for samples collected in month 1 of monitoring, are presented in Figure 9. The array sensor of Pasco VIS-nearIR spectrometer allowed us to record the spectra in seconds, therefore, the sedimentation and/or flotation process will not affect the spectral amplitude function of wavelength. This was a problem when a similar measurement was attempted using a step-by-step UV-VIS spectrometer, when the measurement is performed by gradually changing the wavelength, and then the measurement lasted for several minutes. All wastewater and sludge solution present a large absorption in violet light (possible extended in the ultraviolet domain). One can also observe a certain absorption at the wavelength of ~420 nm, then the discharged wastewater, chemically and biologically treated wastewater present a certain decay of the absorbance with the increasing the light wavelength. For these samples, one can observe a certain absorption doublet in the infrared domain at wavelengths larger than 900 nm. For the untreated wastewater, one can observe an additional relative maximum of absorption at λ~560 nm at a so called Chartreuse color between green and yellow. Similar features are observed in sludge solution and in untreated wastewater, except for the fact that the absorbance through the sludge solution is not decaying for wavelengths larger than 560 nm up to ~840 nm that belong to the near infrared domain.

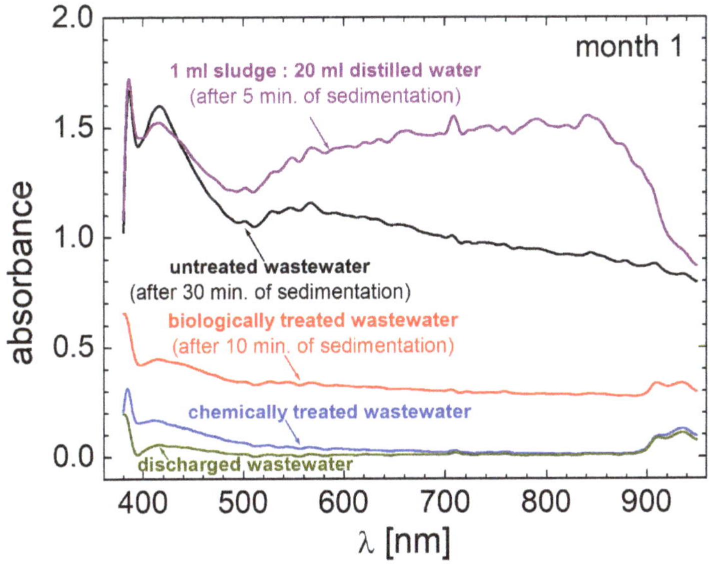

Figure 9. The VIS-nearIR absorbance measured for the chicken slaughterhouse wastewater and 1 mL of sludge residues distilled in 20 mL of distilled water collected in month 1 of monitoring.

An extensive analysis can be performed of the light absorption function of wavelength correlated with the particles sizes and types of pollutants. Such analysis is beyond our scope, therefore will not be discussed here. Nevertheless, it is important to observe that the VIS-nearIR spectroscopy is sensitive to the degree of pollution of wastewater, presenting the largest absorbance for the sludge (even if it was diluted 1:20 in distillated water) and then for the untreated wastewater. Next, less absorbance was measured for the biologically treated wastewater (since it was collected before sedimentation). A small absorbance was measured for chemically treated wastewater and the cleanest wastewater was found the discharged wastewater. Among of all our methods, it is the VIS-nearIR spectroscopy that more accurately reveals the purification process. Unfortunately, this method also can be affected by large errors, as moving particles enter (by sedimentation and/or flotation) between the LED light source and detector, and by data sampling. Going back to a single parameter aimed to characterize the VIS-nearIR spectra, we found the total absorbance, i.e., the integral area under each spectrum. This parameter (in arbitrary units) is presented in Figure 10a for all wastewater samples and sludge and for all four months of monitoring. Large variations can be observed in untreated wastewater, biologically treated wastewater and sludge. The statistical average of the total absorbance and the measurement errors are presented in Figure 10b. Large error bars are observed for the samples enumerated before. One can remark the small values for the averaged total absorbance measured in chemically treated wastewater and in discharged wastewater. This is a new proof that the wastewater purification process applied at the chicken slaughterhouse is efficient. We also performed a correlation/calibration measurement for some milky-like samples over a wide range of turbidity degrees, and we found that the total absorbance measured from VIS-nearIR spectra can replace the turbidity measurement with a proper calibration, since the dependence in the domain 0–500 ntu is linear. Over this limit of 500 ntu, there is a non-linear relationship. The resulted calibration curve is given in Figure S4 in Supplementary Information. Finally, the large variation of data in VIS-nearIR spectra can be reduced by multiple sampling and an appropriate statistical average.

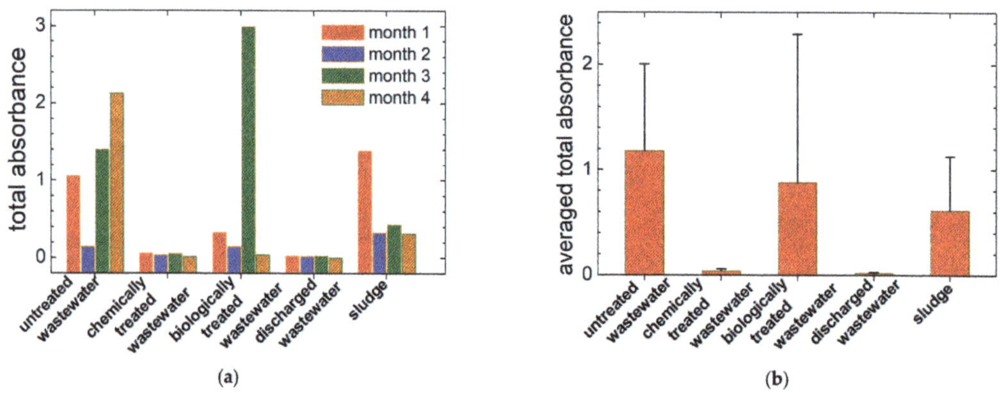

Figure 10. (**a**) The total absorbance of VIS-nearIR spectra recorded for the slaughterhouse wastewater (untreated, chemically treated, biologically treated and discharged water) and sludge collected in all four months of monitoring; (**b**) The averaged of total absorbance of VIS-nearIR spectra of wastewater and sludge presented in (**a**).

3.4. The pH Measurements

A widely used parameter for water testing is its pH. The wastewater pH values for all samples (untreated, chemically and biologically treated, discharged wastewater and sludge solution) collected in all four months were measured and are shown in Figure 11a. Large variations can be observed among samples collected in different months. One can remark a minimum value for pH of ~5.1 and a maximum value of ~8.4, both being measured for samples collected in month 1 of monitoring. Interesting results are obtained for the average values calculated for all four months of monitoring. These values are presented in Figure 11b together with the statistical measurement errors. The average pH values can be found between ~6.5 for discharged wastewater and ~6.8 for chemically and biologically treated wastewater. Nevertheless, within the experimental error limits (the maximum pH error was calculated for chemically treated wastewater at approximately ±1), one cannot distinguish the samples by measuring the pH parameter. Even the sludge solution (the sludge pH was measured for the solution prepared for the VIS-nearIR spectroscopy) has a pH between the same limits. One also observes that the pH was a controlled parameter for the chemically treated wastewater, and its value measured for the wastewater automatically opens the valves of the coagulant, NaOH and liquid polymer tanks.

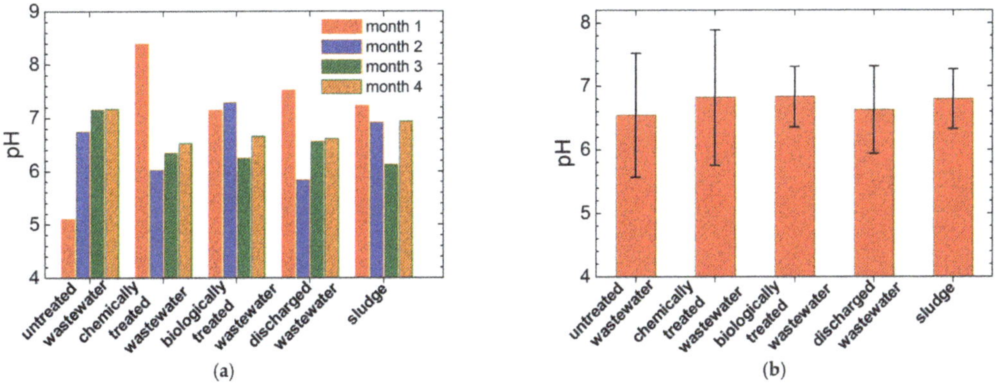

Figure 11. (**a**) The pH measured for the slaughterhouse wastewater (untreated, chemically treated, biologically treated and discharged wastewater) and sludge collected in all four months of monitoring; (**b**) The averaged pH over months measured for all four wastewater and sludge samples presented in (**a**).

3.5. Electrical Conductivity Measurements

The dissolved and undissolved pollutants will change the electrical conductivity (EC) of slaughterhouse wastewater. Figure 12a presents the measured values for all types of wastewater and sludge (in solution as was prepared for VIS-nearIR spectroscopy) collected in all four months of monitoring. As in the case of pH measurements, large differences of the electrical conductivity values can be observed when comparing samples collected in different months. (The average values together with the standard deviations are presented in Figure 12b). In general, for the same type of dissolved particles, a large value measured for the electrical conductivity implies a large concentration of pollutants. The average values of electrical conductivity σ show a decay when comparing untreated wastewater to chemically and biologically treated wastewater. This showed similar values of σ, while the discharged wastewater is characterized by the smallest σ value. Contrary to pH, in the case of electrical conductivity, the value measured for sludge is substantially larger compared to the values measured for all the types of wastewater. Examining the magnitude of error bars, the first idea is that the electrical conductivity, despite the average values, cannot be used to discriminate between different kinds of wastewater. On the contrary, as we will see in the subchapter dedicated to statistical analysis of principal components, not only can the electrical conductivity be successfully used in differentiation between all four types of wastewater but. together with total dissolved solids, it proves to be an important parameter.

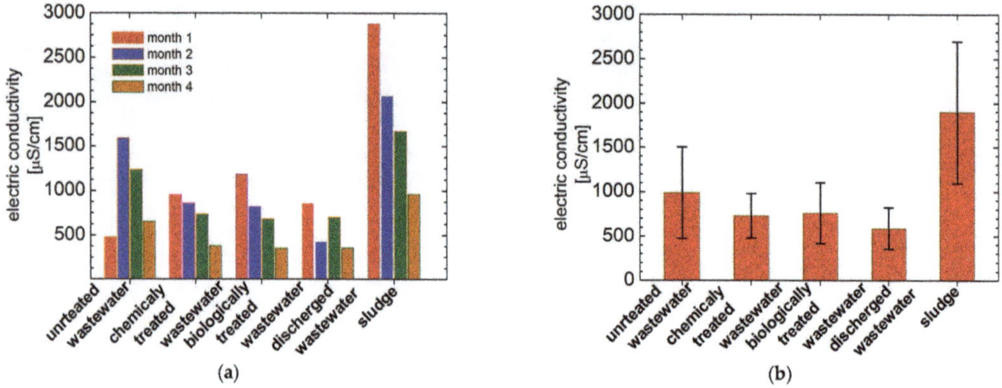

Figure 12. (**a**) The electrical conductivity (EC) measured for the slaughterhouse wastewater (untreated, chemically treated, biologically treated and discharged wastewater) and sludge collected in all four months of monitoring; (**b**) the average EC over months measured for all wastewater and sludge presented in (**a**).

3.6. Total Dissolved Solids' Measurements

A total dissolved solid is a global parameter that, as in the case of electrical conductivity, is usually applied for the characterization of the pollution degree in wastewaters, in particular for those collected from chicken slaughterhouses. In fact, often these two parameters are well correlated. This is a feature that was also observed for our samples. In Figure 13a, the measurement of TDSS values are presented for all the wastewater samples (untreated, chemically and biologically treated and discharged wastewater) and sludge, for all four months of monitoring. With some minor differences, these values look similar to those measured for the electrical conductivity (see Figures 12a and 13a). As in the case of electrical conductivity, the TDSS values are out of range for samples collected in month 1 of monitoring for untreated wastewater, biologically treated wastewater and sludge. Contrary to the measurement of EC in the case of TDSS for month 4 of monitoring, the data seem to be in the same range with the other values measured for the samples collected earlier. The average TDSS values and the measurement errors are comparatively presented in Figure 13b. A large value is obtained for the untreated wastewater (~585 ppm). Similar values are obtained for chemically and biologically treated wastewater (427.75 ppm and 426.5 ppm, respectively). As a result of purification, the discharged wastewater has an average value of TDSS of ~392.5 ppm. The relative elevated value of TDSS measured for the discharged wastewater is in agreement with the ^{1}H NMR diffusometry measurements (Figures 6b and 7), VIS-nearIR spectroscopy (Figure 9), and electrical conductivity (Figure 10), which all showed that the discharged wastewater is not pure, contrary to the interpretation of ^{1}H NMR relaxometry data (Figures 2a and 4). Moreover, it will be shown later than the TDSS measurements are important data for discrimination between different types of wastewater and sludge via statistical analysis of the principal components.

3.7. Principal Component Analysis

The method by which one can analyze variate data types characterized by variables with different measurement units is that of statistical multivariate analysis, in particular the analysis of the principal components [53,54]. For our analysis, we implemented numerically our own analysis program written in MatLab and we plotted the results using Microsoft Excel. The data were verified using PAST Version 3.25 which stands for PAleontological STatistics software and Origin 2022b (Academic). For the principal component analysis we produced an input data matrix with the most relevant parameters (a total of six), as listed in Table 1. As variables, we selected: (i) the total absorbance as measured from VIS-nearIR spectra (see Figure 9); (ii) the $T_{2,1}$ spin–spin relaxation time values measured as the most probable value for the water with dissolved solids (see Figures 2b and 4); (iii) the self-diffusion coefficient D_1 values corresponding to bulk water measured as the most probable coefficient from the maximum of D-distribution (see Figure 6b); (iv) the pH (see Figure 11a); (v) the electrical conductivity, EC (see Figure 12a) and (vi) the total dissolved solids (see Figure 13a). These values are listed for all types of wastewater (untreated, chemically treated, biologically treated and discharged wastewater) and sludge which are considered to represent a particular group, and all four months of monitoring are considered as particular variations in each group. In total, the matrix contains 6 columns (representing the independent parameters) and 20 rows (representing the number of measured data for each parameter). Since the number of independent parameters is smaller than the number of data items, then this will give the number of principal components (PCs). Therefore, in our case the number of PCs was six. The analysis is similar to that in an eigenvectors/eigenvalues problem. The proportion with the contribution of each component and the cumulative proportion are presented in Figure S5a from the Supplementary Information. One can see that the first component (PC1) can explain ~48.9 of the variation in the experimental data. If the PC2 is also considered, then ~69.3% of the variations can be explained. In order to pass over 90%, PC3 and PC4 must also be considered. The eigenvalue functions of the PC number are presented in Figure S5 from the Supplementary Information.

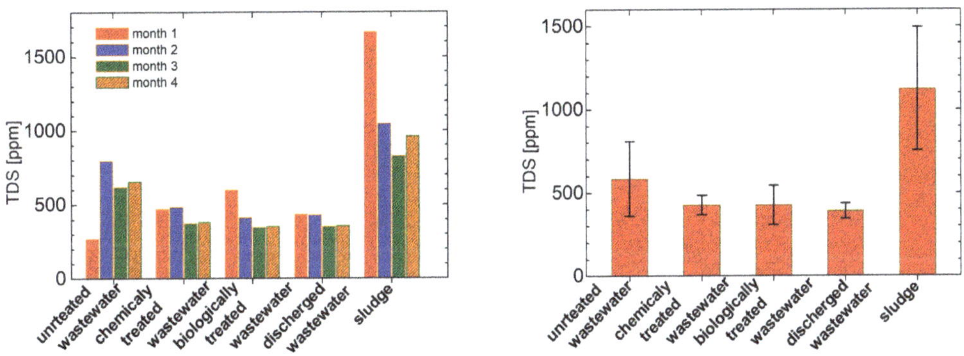

Figure 13. (**a**) The total dissolved solid (TDSS) measured for the slaughterhouse wastewater (untreated, chemically treated, biologically treated and discharged wastewater) and sludge samples collected in all four months of monitoring; (**b**) the average TDSS measured for all wastewater and sludge samples presented in (**a**).

4. Discussion

Principal component analysis is a powerful tool to assess the statistical behavior of measured data, allowing discrimination between different parts of a statistical system, to find correlations between different types of parameters and, if relevant, to observe an evolution process [54]. The main result of the application of PCA analysis on our slaughterhouse wastewater measurements is presented in Figure 14 as a plot of PC2 (second principal component) in correspondence with PC1 (the first principal component). The data must be analyzed in close relationship with the results listed in Table 2. Here, the contributions of each parameter (the total absorbance of VIS-nearIR spectra, $T_{2,1}$ spin–spin relaxation time, the self-diffusion coefficient D_1, pH, EC and TDSS) to the principal components PC1 to PC6 are given. To assess the importance of a specific parameter for each principal component, only the magnitudes of listed values must be considered. The sign of listed values will contribute to the displacement of data presented in Figure 14 and will lead to the separation between different types (groups) of samples. The first principal component PC1 is characterized by the largest separation of data, and the displacement (variation) decays with the increase of principal component number, PC6 having the smallest displacement.

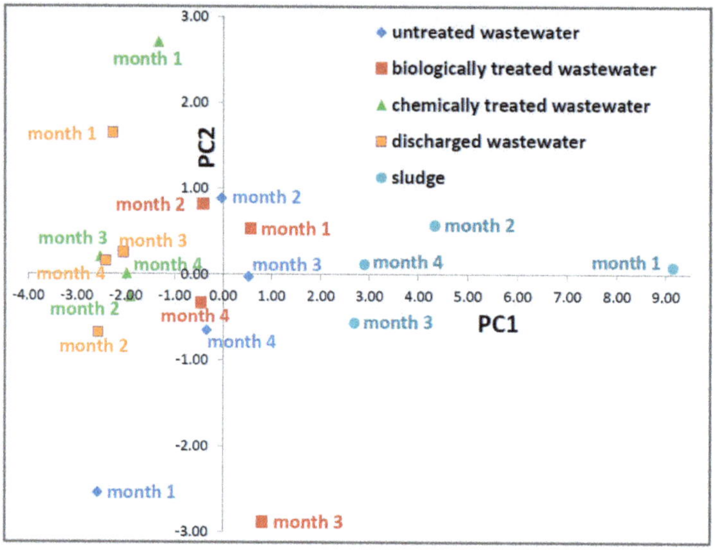

Figure 14. The main PCA analysis: the 2D plot of PC1 function of PC2 for the slaughterhouse wastewater (untreated, biologically treated, chemically treated and discharged wastewater) and sludge samples collected in all four months of monitoring.

Table 2. The contribution of relevant parameters (total VIS-nearIR absorbance, the most probable transverse relaxation time $T_{2,1}$, the self-diffusion coefficient D_1, pH, electrical conductivity EC and the total dissolved solids TDSS) measured for the slaughterhouse wastewater to the principal components PC1–PC6.

Parameter	PC1	PC2	PC3	PC4	PC5	PC6
total VIS-nearIR absorbance	0.336	−0.627	0.550	0.430	−0.079	−0.025
$T_{2,1}$ [s]	−0.561	0.054	0.703	−0.417	0.117	0.026
D_1 [10^{-9} m^2/s]	−0.892	0.343	0.122	0.036	−0.244	−0.104
pH	0.250	0.794	0.264	0.478	0.085	0.032
EC [μS/cm]	0.884	0.220	0.173	−0.288	−0.211	0.118
TDSS [ppm]	0.933	0.165	0.107	−0.226	0.042	−0.195

Note: In all cases, the measurement errors are smaller than 5%.

A PCA statistical analysis is considered to be successful (the number and the type of chosen parameters is relevant, the data are correlated, etc.) if the data presented in a two- (as in the case of the plot in Figure 14) or a multi-dimensional space presents a cluster behavior [49,50]. This means that the points representing the data from the same group are found together and as far away as possible from other groups. Analyzing real data, affected by many types of errors, one can find overlapping regions between different groups and often there will be point(s) that are missed (located away) from a designated group.

In Figure 14, the group of four points (month 1 to four of monitoring) corresponding to untreated wastewater is shown with a blue diamond. For months 2, 3 and 4, these points are located towards the center of the figure having the smallest PC1 and PC2 values. The point corresponding to month 1 of monitoring for untreated wastewater is located at a large negative PC1 and PC2. Table 2 summarizes the main contributions to PC1: the global parameters TDSS (0.933) and EC (0.884), and the specific NMR self-diffusion coefficient

D_1 (−0.892). A certain contribution can be found from the $T_{2,1}$ NMR parameter (−0.561). The smallest contribution to PC1 comes from pH (0.25) and total absorbance in VIS and near IR (0.336). Conversely, the main contribution to the second principal component PC2 comes from pH (0.794). In addition, a large contribution can be credited the total VIS-nearIR absorbance (−0.627). To PC2, an insignificant contribution comes from $T_{2,1}$ (0.054), followed by TDSS (0.165) and EC (0.22). The NMR self-diffusion coefficient D_1 will make a relatively important contribution (0.343). Therefore, the measurements of pH, EC and TDSS performed for samples collected in month 1 of monitoring are out of range. A similar behavior can be found for biologically treated wastewater (red filled square in Figure 14). Here, the points corresponding to samples collected in months 1, 2 and 4 are grouped and the point corresponding to month 3 is largely displaced in the negative direction of PC2. Therefore, the pH value and the total absorbance in VIS-nearIR measured for this sample are the main factors responsible for this out of group position. Grouped at negative values of PC1 can be found the points belonging to chemically treated wastewater (shown with filled green triangles) and discharged wastewater (filled yellow square). Especially in PC1, one can observe a good grouping behavior with only a small variation. From the point of view of PC1, the biologically treated wastewater (PC1: from −0.46 to 0.81—without month 3) is the most similar to the untreated wastewater (PC1: from −2.53 to −1.92—without month 1), while the discharged wastewater (PC1: from −2.57 to −2.06) is the most different from untreated wastewater. From the point of view of the purification process, this is a very good observation. One can extend this kind of analysis. By taking a look at the cluster of discharged wastewater, one can see that the chemically treated wastewater (for months 2, 3 and 4) will have similar properties to those of the discharged wastewater. From the point of view of PC1, the chemically treated wastewater (PC1: from −2.53 to −1.92—without month 1) collected in month 3 of monitoring is cleaner (different from untreated wastewater) than the discharged wastewater collected in months 3, 1 and 4. However, if the data are well grouped in PC1, the points representing the properties of samples collected in month 1 of monitoring for chemically treated and discharged wastewater, in PC2, are away from the group center of gravity. For the most part, the pH measured for these samples (of chemically treated wastewater and discharged wastewater) led to this displacement.

A completely opposed behavior is observed for sludge (blue circles). The group with these samples is localized at positive large values of PC1, well separated even from untreated wastewater. In the second component, one observes that the values are distributed around the zero value of PC2. Then, one can conclude that PC1 can be used to successfully separate the wastewater according to is purity. The most contaminated sample was the sludge sample, collected in month 1 of monitoring, and the cleanest wastewater was the wastewater sample collected in month 2 of monitoring. Another study based on FT-IR spectroscopy supports this conclusion [9]. For our wastewater system (also containing the sludge), one can see that the most significant parameters were (in order): (i) the total dissolved solids TDSS; (ii) the self-diffusion coefficient D_1; (iii) the electrical conductivity EC; (iv) the transverse relaxation time of dissolved solids $T_{2,1}$. The most irrelevant parameter was pH. Nevertheless, pH and the total absorbance in VIS-nearIR are important parameters capable of identifying samples (in PC2) from the same group with properties away from the average expectation.

Another important result of our PCA assay is the degree of correlation between all parameters considered in pairs. In the case of wastewater collected from a chicken slaughterhouse, the most correlated parameters were found to be the electrical conductivity, EC, and the total dissolved solids, TDSS. This is not a surprise, since the physical support for electrical conductivity is given by dissolved solids, as distilled water presents poor electrical conductivity. Considering the elements of covariance matrix C (using the MatLab 2020 numeric program, but not from PAST software), and that the eigenvector and eigenvalue problems were solved, a large value can be seen for the elements (there are two symmetrical elements around a main diagonal that have values of 1, meaning that each parameter is perfectly correlated with itself) corresponding to the EC and TDSS parameters. However, the values are not as large as expected. To understand this, in Figure 15, the TDSS is plotted as a function of EC. Now, the correlation coefficient can be better understood. The first observation is that, indeed, the TDSS is well correlated to the EC parameter but, in our case, for all samples collected from the chicken slaughterhouse in four months of monitoring, there are two linear dependences rather than a single one. With one exception, one can see that the data (TDSS and EC) measured for months 1, 2 and 3 of monitoring for untreated, chemically and biologically treated wastewater discharged wastewater and sludge are in a very good linear correlation. In addition, the data collected in the four months of monitoring are arranged in a second linear dependence. Usually, such double linear correlations can be explained: (i) by a change in the treatment process (to include different types of particles with a different electrical behavior); (ii) by a change in dissolved pollutants; (iii) by both causes. In our case, the most probable cause is the fact that the wastewater samples collected in month 4 of monitoring of wastewater were used in the processing of different kinds of chicken (with different origin, size, age, etc.) An interesting observation is the fact that the point corresponding to the discharged wastewater collected in month 2 is placed on the short dependency belonging to month 4.

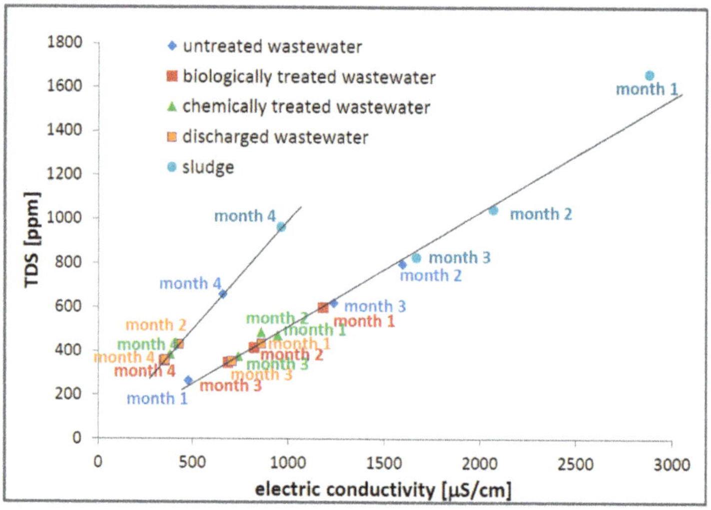

Figure 15. The double linear correlation between electrical conductivity (EC) and totally dissolved solids (TDSs) for the wastewater and sludge.

In the recent times, artificial intelligence is used in more and more fields, from entering parking places to cancer prediction [54]. This is useful, especially in cases where the classes/categories are not well isolated, superposition between them exists and a probabilistic based decision has to be considered. In this sense, we developed a dedicated program written in JavaScript (JS–version 2024) that used an online library named ML5 [55,56] to easily build an Artificial Neural Network. This is used to predict the probability that a new measurement of all parameters analyzed by PCA will be classified as untreated wastewater, chemically or biologically treated wastewater, discharged wastewater, or sludge (see Figure 16). We simulated the incidence of such analysis in every element of a 100×100 matrix associated with the 2D plot of the main PCA analysis (see Figure 14). Prior to that, the JS loaded the $5 \times 4 = 20$ values of PC1 and PC2 coordinated, together with the label associated with our five types of samples (four wastewater and sludge). The ANN was trained for 10,000 epochs with a learning rate of 10% and the trained model was saved to be loaded anytime needed (without any new training, which is a time-consuming procedure). Once the trained model is loaded, it can be used to generate prediction related to the probability that, in a chosen position (in the main analysis 2D space) resulting from a PCA analysis of a new element, the sample is associated with one of the trained labels (i.e., untreated wastewater, chemically treated wastewater, biologically treated wastewater, evacuated wastewater or sludge). The generated prediction maps, which simplify the PCA analysis even further, are presented in Figure 16. Thus, Figure 16a presents the predicted probability (measured from 0 to 1, where 1 represents 100%) for untreated wastewater. One can observe that the maximum probability is obtained for PC1 and PC2 negative (see also Figure 14), especially due to the sample collected in month 1. A relative maximum (around 50%) is obtained for PC1 and PC2 values close to zero. especially due to the samples collected in months 2 and 3. The PCA analysis of chemically treated wastewater can be found with high probability (>70%) if PC1 is between −3 to −1 and PC2 is larger than 1 (see Figure 16c). A high probability of association, in the PCA analysis, between the measured result with biologically treated wastewater is to have PCA between −1 and 0, and PC2 with large positive or negative values (see Figure 16c). From Figure 16c, one can see that a smaller PC1 value (almost independent of PC2 value) is associated with the discharged wastewater (see Figure 16d). Values of PC1 larger than zero can be associated with high probability to measured sludge (see Figure 16e). The prediction maps can offer enhanced representation of sample classification related to the PCA analysis.

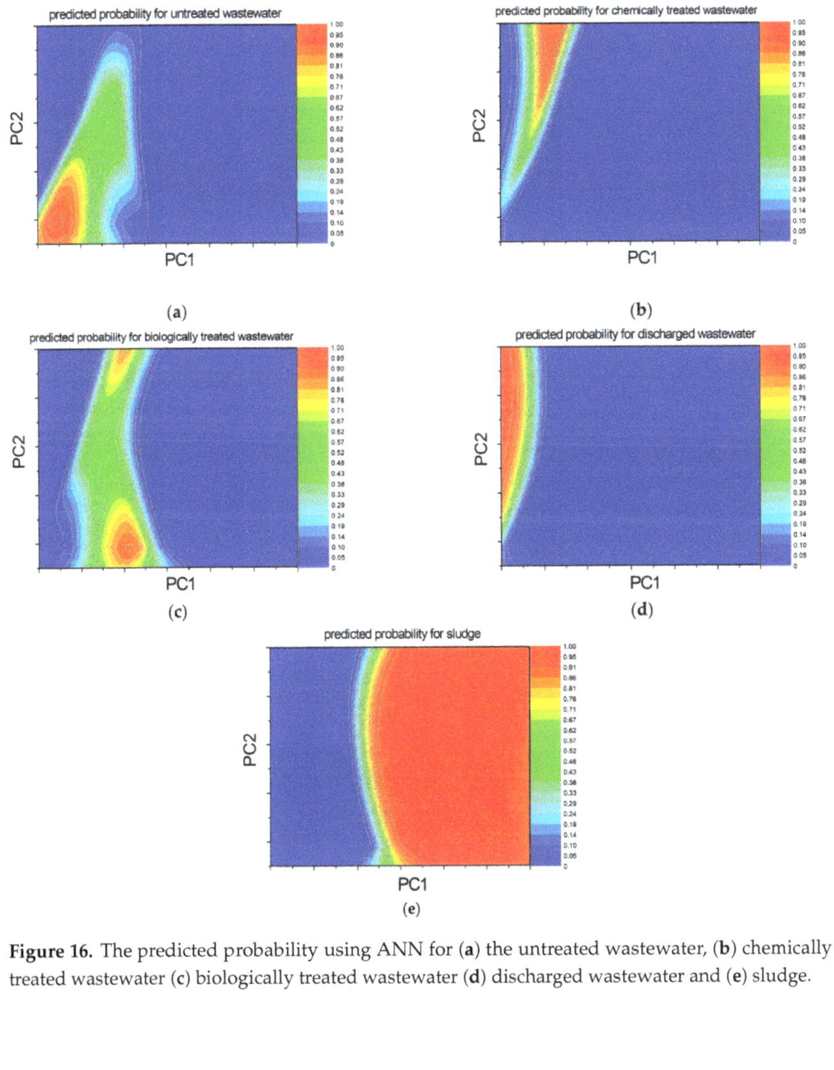

Figure 16. The predicted probability using ANN for (**a**) the untreated wastewater, (**b**) chemically treated wastewater (**c**) biologically treated wastewater (**d**) discharged wastewater and (**e**) sludge.

Such prediction maps were produced and discussed in [54], where Dragan et al. classify data measured by NMR and FT-IR to various types of colorectal cancer. However, according to our knowledge, this is the first time that such predictions are used to classify the wastewater and sludge resulted from slaughterhouse activity. Thus, one can say that this predictive approach validates the experimental data, which may be used an automated real-time monitoring along with an educated decision-making processes.

5. Conclusions

This is one of the most complex studies related to the characterization of wastewater collected from slaughterhouses. This complexity is based on both the large variety of measured parameters using modern and classical methods, and on the many levels of analysis which start with raw measurement, then continue with the ILT analysis of measured data and primary classification. This then continues with statistical analysis of principal components, which is finally used to train an artificial neural network to predict and classify new measurements. The advantage was demonstrated of using model-free low field ^{1}H NMR relaxometry and diffusometry combined with Laplace-like analysis compared to the classical methods, which provide a global single value for the entire sample, leading to a more realistic characterization of wastewater dynamic components. This analysis showed that the distributions of ^{1}H NMR transverse relaxation times T_2 present two distinct domains. The first characterizes water with dissolved solids and the second characterizes water with undissolved solids. Additionally, it was shown that the distribution of ^{1}H NMR self-diffusion coefficient D also presents two domains. These are self-diffusion in bulk water and the transported water attached to solid residues flowing by flotation and sedimentation. The ^{1}H NMR diffusometry measurements proved this, even though the discharged wastewater is not completely clean of undissolved solids. To assess the contribution of each parameter to the evaluation process of the efficiency of the purification treatment performed at a chicken slaughterhouse, a statistical analysis of principal components was performed. It was shown how the PCA analysis can be further improved if it is used to train an ANN to predict and classify a new measurement for one of the studied categories of wastewater or sludge. From the analysis of the distributions of transverse relaxation times T_2 and diffusion coefficient D, the VIS-nearIR spectra and EC and TDSS measured parameters, one can conclude that the purification process performed by the chicken slaughterhouse is efficient. Finally, this study demonstrates that, in addition to classical methods of analysis, which have their own merits but are limited to a single-value global characterization, the use of complementary methods, capable of emphasizing different components by providing a distribution of values for measured parameters and multi-level analysis, is an extraordinary tool that shows how each technique used for wastewater treatment works, what its advantages and disadvantages are, what its efficiency is, and what its limitations are and, above all, how it can be improved for a better treatment strategy. In particular, the proposed complex set of measurements of wastewater properties at every stage of the treatment process can provide useful information related to the complex interaction between treatment agent and wastewater components. Theis type of information can be used to design improved strategies for wastewater purification. The complex analysis (PCA combined with machine learning capable of prediction) can be used by industry to automate the monitoring process of wastewater treatment, but it is not limited to this. All these presented methods, experimental and numerical, are easily implemented in worldwide laboratories, since both low-field NMR and VIS-nearIR spectrometers have a relatively low acquisition and exploitation cost, therefore presenting a high probability of becoming a new standard in the field.

Supplementary Materials: The following are available online at https://www.mdpi.com/article/10.3390/w16172382/s1, Figure S1: The CPMG and PGSE pulse sequences; Figure S2: The T_2-distributions measured for sludge; Figure S3: The self-diffusion coefficient D distributions measured for sludge; Figure S4: The relationship between the total absorbance in the VIS-NearIR spectra and turbidity and Figure S5: The proportions and eigenvalues resulted from PCA applied on the measured parameters of poultry slaughterhouse wastewater.

Author Contributions: R.C. collected the samples, conceived, designed and performed the experiments, primary and secondary data analysis and statistical analysis, revised and proofread the paper; R.F. conceived and designed the experiments, wrote the numerical software for advanced data analysis, wrote and revised the paper. All authors have read and agreed to the published version of the manuscript.

Funding: This research received no external funding.

Data Availability Statement: The original contributions presented in the study are included in the article/Supplementary Materials, further inquiries can be directed to the corresponding author/s.

Acknowledgments: The authors would like to thank S.C. AGROPROD CRASNA Cooperativa Agricolă S.R.L., Satu-Mare, Romania, in particular Daniel Pereș and Petru Crainic. for providing the chicken slaughterhouse wastewater samples and relevant information about the wastewater treatment. The authors thank also Simona Nicoara for suggestions for English corrections.

Conflicts of Interest: The authors declare no conflicts of interest.

References

1. Paulista, L.O.; Presumido, P.H.; Peruço Theodoro, J.D.; Novaes Pinheiro, A.L. Efficiency analysis of the electrocoagulation and electroflotation treatment of poultry slaughterhouse wastewater using aluminum and graphite anodes. *Environ. Sci. Poll. Res.* **2018**, *25*, 19790–19800. [CrossRef]
2. Aceves-Lara, C.-A.; Latrille, E.; Conte, T.; Steye, J.-P. Online estimation of VFA, alkalinity and bicarbonateconcentrations by electrical conductivity measurementduring anaerobic fermentation. *Water Sci. Technol.* **2012**, *65*, 1281–1289. [CrossRef]
3. Yusoff, M.S.; Azwan, A.M.; Zamri, M.F.M.A.; Aziz, H.A. Removal of colour, turbidity, oil and grease for slaughterhouse wastewater using electrocoagulation method. *AIP Conf. Proc.* **2017**, *1892*, 040012.
4. Khanna, V.K. pH measurement of dirty water sources by ISFET: Addressing practical problems. *Sens. Rev.* **2007**, *27*, 233–238. [CrossRef]
5. Modla, M. The easy guide to pH measurement. *Meas. Control* **2004**, *37*, 204–206. [CrossRef]
6. Ferraz, F.M.; Yuan, Q. Nitrite interference with soluble COD measurements from aerobically treated wastewater. *Water Environ. Res.* **2017**, *89*, 549–554. [CrossRef]
7. APHA. *Standard Methods for the Examination of Water and Wastewater*, 22nd ed.; American Public Health Association: Washington, DC, USA, 2012.
8. Maremane, S.; Belle, G.; Oberholster, P. Assessment of Effluent Wastewater Quality and the Application of an Integrated Wastewater Resource Recovery Model: The Burgersfort Wastewater Resource Recovery Case Study. *Water* **2024**, *16*, 608. [CrossRef]
9. Alekseevsky, D.; Chernysh, Y.; Shtepa, V.; Chubur, V.; Stejskalová, L.; Balintova, M.; Fukui, M.; Roubík, H. Enhancing Ecological Efficiency in Biological Wastewater Treatment: A Case Study on Quality Control Information System. *Water* **2023**, *15*, 3744. [CrossRef]
10. El Aatik, A.; Navarro, J.M.; Martínez, R.; Vela, N. Estimation of Global Water Quality in Four Municipal Wastewater Treatment Plants over Time Based on Statistical Methods. *Water* **2023**, *15*, 1520. [CrossRef]
11. Mamun, M.; Kim, J.Y.; An, K.G. Multivariate Statistical Analysis of Water Quality and Trophic State in an Artificial Dam Reservoir. *Water* **2021**, *13*, 186. [CrossRef]
12. Boughou, N.; Majdy, I.; Cherkaoui, E.; Khamar, M.; Nounah, A. Physico-chemical characterization of wastewater from slaughterhouse: Case of rabat in Morocco. *ARPN J. Agric. Biol. Sci.* **2018**, *13*, 19–24.
13. Crainic, R.; Drăgan, L.R.; Fechete, R. ^{1}H NMR relaxometry and ATR-FT-IR spectrosopy used for the assesment of wastewater treatment in slaughterhouse. *Studia UBB Phys.* **2018**, *63*, 49–60. [CrossRef]
14. Seif, H.; Moursy, A. Treatment of slaughterhouse wastes. In Proceedings of the Sixth International Water Conference, IWTC, Alexandria, Egypt, 23–25 March 2001; pp. 270–271.
15. Alayu, E.; Yirgu, Z. Advanced technologies for the treatment of wastewaters from agro-processing industries and cogeneration of by-products: A case of slaughterhouse, dairy and beverage industries. *Int. J. Environ. Sci. Technol.* **2018**, *15*, 1581–1596. [CrossRef]
16. López-Maldonado, E.A.; Oropeza-Guzman, M.T.; Jurado-Baizaval, J.L.; Ochoa-Terán, A. Coagulation–flocculation mechanisms in wastewater treatment plants through zeta potential measurements. *J. Hazard. Mat.* **2014**, *279*, 1–10. [CrossRef]

17. Mehmood, K.; Rehman, S.K.U.; Wang, J.; Farooq, F.; Mahmood, Q.; Jadoon, A.M.; Javed, M.F.; Ahmad, I. Treatment of Pulp and Paper Industrial Effluent Using Physicochemical Process for Recycling. *Water* **2019**, *11*, 2393. [CrossRef]
18. Zamani, H.; Golestani, H.A.; Mousavi, S.M.; Zhiani, R.; Hosseini, M.S. Slaughterhouse wastewater treatment using biological anaerobic and coagulation-flocculation hybrid process. *Desal. Wat. Treat.* **2019**, *155*, 64–71. [CrossRef]
19. Lazaridis, N.K.; Matis, K.A.; Webb, M. Flotation of metal-loaded clayanion exchangers. Part I: The case of chromate. *Chemosphere* **2001**, *42*, 373–378. [CrossRef]
20. Doyle, F.M.; Liu, Z.D. The effect of triethylenetraamine (trien) on the ion flotation of Cu^{2+} and Ni^{2+}. *J. Colloid Interface Sci.* **2003**, *258*, 396–403. [CrossRef]
21. Khandegar, V.; Saroha, A.K. Electrocoagulation for the treatment of textile industry effluent e A review. *J. Environ. Manag.* **2013**, *128*, 949–963. [CrossRef]
22. Ozturk, D.; Yilmaz, A.E. Treatment of slaughterhouse wastewater with the electrochemical oxidation process: Role of operating parameters on treatment efficiency and energy consumption. *J. Water Proc. Eng.* **2019**, *31*, 100834. [CrossRef]
23. Tünay, O.; Kabdasli, N.I. Hydroxide precipitation of complexed metals. *Water Res.* **1994**, *28*, 2117–2124. [CrossRef]
24. Andrus, M.E. A review of metal precipitation chemicals for metal-finishing applications. *Metal Finish.* **2000**, *98*, 20–23. [CrossRef]
25. Chen, B.; Qu, R.; Shi, J.; Li, D.; Wei, Z.; Yang, X.; Wang, Z. Heavy metal and phosphorus removal from water by optimizing use of calcium hydroxide and risk assessment. *Environ. Pollut.* **2012**, *1*, 38–54. [CrossRef]
26. Yang, Z.; Zhou, Y.; Feng, Z.; Rui, X.; Zhang, T.; Zhang, Z. A Review on Reverse Osmosis and Nanofiltration Membranes for Water Purification. *Polymers* **2019**, *11*, 1252. [CrossRef] [PubMed]
27. Hacıfazlıoğlu, M.C.; Parlar, I.; Pek, T.Ö.; Kabay, N. Evaluation of chemical cleaning to control fouling on nanofiltration and reverse osmosis membranes after desalination of MBR effluent. *Desalination* **2019**, *466*, 44–51. [CrossRef]
28. Aziz, H.A.; Puat, N.N.A.; Alazaiza, M.Y.D.; Hung, Y.T. Poultry Slaughterhouse Wastewater Treatment Using Submerged Fibers in an Attached Growth Sequential Batch Reactor. *Int. J. Environ. Res. Public Health* **2018**, *15*, 1734. [CrossRef]
29. Pan, M.; Huang, X.; Wu, G.; Hu, Y.; Yang, Y.; Zhan, X. Performance of Denitrifying Phosphate Removal via Nitrite from Slaughterhouse Wastewater Treatment at Low Temperature. *Water* **2017**, *9*, 818. [CrossRef]
30. Álvarez-Méndez, S.J.; Ramos-Suárez, J.L.; Ritter, A.; González, J.M.; Pérez, Á.C. Anaerobic digestion of commercial PLA and PBAT biodegradable plastic bags: Potential biogas production and ^{1}H NMR and ATR-FTIR assessed biodegradation. *Heliyon* **2023**, *9*, e16691. [CrossRef]
31. Stapf, S.; Kimmich, R. Translational versus rotational molecular dynamics in plastic crystals studied by NMR relaxometry and diffusometry. *Mol. Phys.* **1997**, *92*, 1051–1060. [CrossRef]
32. Martini, F.; Carignani, E.; Nardelli, F.; Rossi, E.; Borsacchi, S.; Cettolin, M.; Susanna, A.; Geppi, M.; Calucci, L. Glassy and Polymer Dynamics of Elastomers by ^{1}H Field-Cycling NMR Relaxometry: Effects of Cross-Linking. *Macromolecules* **2020**, *53*, 10028–10039. [CrossRef]
33. Asano, A. Chapter One-NMR Relaxation Studies of Elastomers. In *Annual Reports on NMR Spectroscopy*; Academic Press: Cambridge, MA, USA, 2015; Volume 86, pp. 1–72. [CrossRef]
34. Wang, N.; Xia, Y. Anisotropic analysis of multi-component T_2 and $T_{1\rho}$ relaxations in achilles tendon by NMR spectroscopy and microscopic MRI. *J. Magn. Reson. Imaging* **2013**, *38*, 625–633. [CrossRef] [PubMed]
35. Navon, G.; Eliav, U.; Demco, D.E.; Blümich, B. Study of order and dynamic processes in tendon by NMR and MRI. *J. Magn. Reson. Imaging* **2007**, *25*, 362–380. [CrossRef] [PubMed]
36. Istrate, D.; Er Rafik, M.; Popescu, C.; Demco, D.E.; Tsarkova, L.; Wortmann, F.-J. Keratin made micro-tubes: The paradoxical thermal behavior of cortex and cuticle. *Int. J. Biol. Mac.* **2016**, *89*, 592–598. [CrossRef]
37. Demco, D.E.; Utiu, L.; Tillmann, W.; Blümich, B.; Popescu, C. Morphology and molecular dynamics of hard α-keratin under pressure by ^{1}H and ^{13}C solid-state NMR. *Chem. Phys. Lett.* **2011**, *509*, 62–66. [CrossRef]
38. Colicchio, I.; Demco, D.E.; Baias, M.; Keul, H.; Moeller, M. Influence of the silica content in SPEEK–silica membranes prepared from the sol–gel process of polyethoxysiloxane: Morphology proton mobility. *J. Membr. Sci.* **2009**, *337*, 125–135. [CrossRef]
39. Baias, M.; Demco, D.E.; Blümich, B.; Möller, M. State of water in hybrid sulfonated poly(ether ether ketone)–silica membranes by ^{1}H solid-state NMR. *Chem. Phys. Lett.* **2009**, *473*, 142–145. [CrossRef]
40. Lacan, I.; Moldovan, M.; Ardelean, I. The Influence of Chitosan on Water Absorption and Solubility of Calcium Phosphate Cement. *Coatings* **2023**, *13*, 1641. [CrossRef]
41. Toma, I.-O.; Stoian, G.; Rusu, M.-M.; Ardelean, I.; Cimpoeşu, N.; Alexa-Stratulat, S.-M. Analysis of Pore Structure in Cement Pastes with Micronized Natural Zeolite. *Materials* **2023**, *16*, 4500. [CrossRef]
42. Thomas, D.; Oros-Peusquens, A.-M.; Poot, D.; Shah, N.J. Whole-Brain Water Content Mapping Using Super-Resolution Reconstruction with MRI Acquisition in 3 Orthogonal Orientations. *Magn. Reson. Med.* **2022**, *88*, 2117–2130. [CrossRef]
43. Gussoni, M.; Greco, F.; Vezzoli, A.; Paleari, M.A.; Moretti, V.M.; Lanza, B.; Zetta, L. Osmotic and aging effects in caviar oocytes throughout water and lipid changes assessed by ^{1}H NMR T_1 and T_2 relaxation and MRI. *Magn. Reson. Imaging* **2007**, *25*, 117–128. [CrossRef]
44. Wu, B.; Zhou, K.; He, Y.; Chai, X.; Dai, X. Unraveling the water states of waste-activated sludge through transverse spin-spin relaxation time of low-field NMR. *Water Res.* **2019**, *155*, 266–274. [CrossRef] [PubMed]

45. Fechete, R.; Morar, I.A.; Moldovan, D.; Chelcea, R.I.; Crainic, R.; Nicoara, S.C. Fourier and Laplace-like low-field NMR spectroscopy: The perspectives of multivariate and artificial neural networks analyses. *J. Magn. Reson.* **2021**, *324*, 106915. [CrossRef] [PubMed]
46. Venkataramanan, L.; Song, Y.Q.; Hürlimann, M.D. Solving Fredholm integrals of the first kind with tensor product structure in 2 and 2.5 dimensions. *IEEE Trans. Sign. Proc.* **2002**, *50*, 1017–1026. [CrossRef]
47. Song, Y.Q.; Venkataramanan, L.; Hürlimann, M.D.; Flaum, M.; Frulla, P.; Straley, C. T_1–T_2 correlation spectra obtained using a fast two-dimensional Laplace inversion. *J. Magn. Reson.* **2002**, *154*, 261–268. [CrossRef]
48. Hürlimann, M.D.; Flaum, M.; Venkataramanan, L.; Flaum, C.; Freedman, R.; Hirasaki, G.J. Diffusion-relaxation distribution functions of sedimentary rocks in different saturation states. *Magn. Reson. Imaging* **2003**, *21*, 305–310. [CrossRef] [PubMed]
49. Fechete, R.; Demco, D.E.; Blümich, B. Parameter maps of ^{1}H residual dipolar couplings in tendon under mechanical load. *J. Magn. Reson.* **2003**, *165*, 9–17. [CrossRef]
50. Aursand, I.G.; Veliyulin, E.; Böcker, U.; Ofstad, R.; Rustad, T.; Erikson, U. Water and salt distribution in Atlantic salmon (Salmo salar) studied by low-field ^{1}H NMR, ^{1}H and ^{23}Na MRI and light microscopy: Effects of raw material quality and brine salting. *J. Agric. Food Chem.* **2008**, *57*, 46–54. [CrossRef]
51. Cheng, S.; Tang, Y.; Zhang, T.; Song, Y.; Wang, X.; Wang, H.; Wang, H.; Tan, M. Approach for monitoring the dynamic states of water in shrimp during drying process with LF-NMR and MRI. *Dry. Technol.* **2018**, *36*, 841–848. [CrossRef]
52. Carneiro, C.D.S.; Mársico, E.T.; Ribeiro, R.D.O.R.; Conte Júnior, C.A.; Álvares, T.S.; De Jesus, E.F.O. Quality attributes in shrimp treated with polyphosphate after thawing and cooking: A study using physicochemical analytical methods and Low-Field ^{1}H NMR. *J. Food Process. Eng.* **2013**, *36*, 492–499. [CrossRef]
53. Afifi, A.; May, S.; Clark, V.A. *Practical Multivariate Analysis*, 5th ed.; CRC Press: Boca Raton, FL, USA; Taylor and Francis Group: New York, NY, USA, 2012; pp. 357–379.
54. Dragan, L.R.; Andras, D.; Fechete, R. Fourier Transform Infrared (FT-IR) Spectroscopy and Proton Nuclear Magnetic Resonance (^{1}H NMR) Relaxometry and Diffusometry for the Identification of Colorectal Cancer in Blood Plasma. *Anal. Lett.* **2023**, *56*, 286–302. [CrossRef]
55. Shiffman, D. The Nature of Code. In *Simulating the Natural Systems with Processing*; No Starch Press: San Francisco, CA, USA, 2012.
56. Shiffman, D. The Nature of Code. In *Simulating the Natural Systems with Javascript*; No Starch Press: San Francisco, CA, USA, 2024.

water

MDPI

Article

The Potential of AOP Pretreatment in the Biodegradation of PS and PVC Microplastics by *Candida parapsilosis*

Kristina Bule Možar [1], Martina Miloloža [1], Viktorija Martinjak [1], Magdalena Ujević Bošnjak [2], Marinko Markić [1], Tomislav Bolanča [1], Matija Cvetnić [1], Dajana Kučić Grgić [1,*] and Šime Ukić [1,*]

[1] Faculty of Chemical Engineering and Technology, University of Zagreb, Trg Marka Marulića 19, 10000 Zagreb, Croatia; kbule@fkit.unizg.hr (K.B.M.); miloloza@fkit.unizg.hr (M.M.); vprevaric@fkit.unizg.hr (V.M.); mmarkic@fkit.unizg.hr (M.M.); tbolanca@fkit.unizg.hr (T.B.); mcvetnic@fkit.unizg.hr (M.C.)

[2] Croatian Institute of Public Health, Rockefellerova 7, 10000 Zagreb, Croatia; magdalena.ujevic@hzjz.hr

* Correspondence: dkucic@fkit.unizg.hr (D.K.G.); sukic@fkit.unizg.hr (Š.U.); Tel.: +385-14597238 (D.K.G.); +385-14597217 (Š.U.)

Abstract: Microplastics are an emerging class of recalcitrant organic pollutants that are of general scientific and public interest nowadays. It would be ideal to remove microplastics from the environment through biodegradation, as biodegradation is a highly ecological and economically acceptable approach. Unfortunately, the efficiency of biodegradation of conventional plastic polymers is low. The application of a suitable pretreatment could increase the efficiency of biodegradation. In this study, the applicability of UV-C/H_2O_2 and UV-C/$S_2O_8^{2-}$ advanced oxidation processes as pretreatments for the biodegradation of polystyrene and poly(vinyl chloride) microplastics by the yeast *Candida parapsilosis* was investigated. For the investigated range (pH 4–10, peroxide concentration up to 20 mM and treatment duration up to 90 min), the UV-C/H_2O_2 process proved to be more effective in degrading polystyrene microplastics, while the UV-C/$S_2O_8^{2-}$ process was more efficient at degrading poly(vinyl chloride) microplastics. Samples pretreated under optimal conditions (90 min treatment time at a pH of 5.7 and H_2O_2 concentration of 20.0 mM for polystyrene samples; 90 min treatment time at a pH of 8.6 and $S_2O_8^{2-}$ concentration of 11.1 mM for poly(vinyl chloride) samples) were subjected to biodegradation by *Candida parapsilosis*. The biodegradation conditions included an agitation speed of 156 rpm and an initial pH of 5.7 for the experiments with the polystyrene samples, while an agitation speed of 136 rpm and an initial pH of 4.9 were used for the poly(vinyl chloride) experiments. The initial value of the optical density of the yeast suspension was 1.0 in both cases. The experiments showed a positive effect of the pretreatment on the number of yeast cells on the surface of the microplastics.

Keywords: microplastics; polystyrene; poly(vinyl chloride); advanced oxidation; UV-C/H_2O_2; UV-C/$S_2O_8^{2-}$

1. Introduction

Plastic is an extremely useful polymer material with incredible application properties [1]. Unfortunately, a large number of plastic products end up in the environment, posing a serious environmental threat [2]. Small plastic particles, known as microplastics (MPs, particles with a size of 1–5000 μm) and nanoplastics (NPs, particles with a size of less than 1 μm), are considered particularly hazardous [3].

The *2030 Agenda for Sustainable Development*, adopted by the United Nations in September 2015 [4], includes 17 Sustainable Development Goals (SDGs), and half of these goals are directly affected by the problem of plastic pollution [5]. As part of SDG 12, *Responsible Consumption and Production*, the agenda promotes a more rational use of raw materials and products. In the context of plastics, this would inevitably lead to less plastic waste entering the environment. Furthermore, it should not be forgotten that the production of conventional plastic polymers (polyethylene, polypropylene, polyethylene terephthalate, polystyrene (PS) and poly(vinyl chloride) (PVC)) is associated with high CO_2 emissions, as these polymers are made from fossil fuels, and their production requires high energy consumption. A reduction in plastic production is therefore linked to SDG *13, Climate Action*. The effective collection and processing of plastic waste would be a step towards achieving SDG *11, Sustainable Cities and Communities*. Conventional plastic polymers are recalcitrant organic pollutants. The presence of their MPs or NPs in the environment affects the achievement of the SDGs in various ways. MP and NP particles can easily enter the animal/human body due to their small size and cause harmful biological effects [6]. Also, do not forget that MPs serve as vectors for pathogens when adsorbed on their surface [7]. Finally, MP and NP particles from commercially available plastics usually contain various additives, some of which have a high toxic potential. As these additives are not covalently bound, they leach unhindered into the environment [8]. All of this has a direct impact on four SDGs: SDG *3, Good Health and Well-being, SDG 4, Clean Water and Sanitation*, SDG *14, Protection of Seas and Oceans* and SDG *15, Repair Ecosystems and Retain Biodiversity*.

In recent decades, scientists have been intensively researching methods to effectively remove MPs and NPs from the environment, especially from the aquatic environment. Some of the conventional approaches to remove MPs from aquatic media are adsorption on a suitable adsorbent and coagulation with a suitable coagulant. However, the addition of adsorbents or coagulants can provide a new source of water pollution. Membrane processes are considered to be very successful in the removal of MPs. The problem with membrane processes, however, is the high operating and maintenance costs. All three processes mentioned have a common problem: the sludge produced during treatment. The sludge, which is rich in MPs, must be treated further [9]. However, it is often used untreated on agricultural land, creating another pathway for MPs in the environment [10,11].

Biotreatment is considered the most environmentally and economically acceptable approach to the remediation of polluted environments [12]. Due to their rapid reproduction rate and high metabolic potential, bacteria and fungi are able to adapt to new substrates and are therefore the most studied organisms for the biodegradation of conventional plastic polymers [13]. The pathway for the biochemical degradation of MPs is not uniform and depends on the polymer treated and the characteristics of the enzymatic system of the microorganism used for the treatment. Before the biodegradation of MPs can begin, several general conditions must be met: Firstly, the surface of the MPs must be such that microorganisms can adhere to the surface [14]; this is the basic prerequisite for the formation of a biofilm on the surface of MPs. Secondly, the selected microorganisms must be able to produce extracellular enzymes that can break down the plastic polymer into smaller molecules. Finally, these microorganisms must also be able to utilize these smaller molecules as a carbon source, i.e., as an energy source, with the help of an intracellular enzyme system [15]. However, the conventional plastic polymers are extremely stable substances whose structures do not have functional groups that are frequently attacked by enzymes [16]. In addition, these polymers are hydrophobic [17], which makes it difficult for most bacteria and fungi to adhere to the surface of the plastic particles. Most bacteria have hydrophilic cell membranes [18], as do yeast cells and the vegetative hyphae of filamentous fungi that grow in humid environments [19]. Therefore, conventional plastics are not readily biodegradable [20], which complicates the biological approach to solving the problem of MPs in the environment. Therefore, it seems reasonable to apply some kind of a pretreatment that would facilitate biodegradation.

One of the interesting approaches is the use of advanced oxidation processes (AOPs). AOPs are based on the generation of highly reactive radicals with pronounced oxidative capabilities. Therefore, these radicals are able to oxidize a wide range of recalcitrant organic pollutants [21]. The first step in the radical degradation of MPs is either the abstraction of hydrogen or the scission of the C-C bond, which leads to alkyl radicals on the surface of the MPs. Further oxidation leads to chain branching and scission, resulting in the formation of oxygen-containing groups (hydroxyl, peroxide and carbonyl groups) and microcracks in the polymer surface [22]. Although the effectiveness of AOPs in removing various recalcitrant organic pollutants is well documented, AOPs are generally non-selective treatments that require the significant consumption of chemicals, and very often, the use of electricity, which calls into question their cost effectiveness [23]. It is estimated that the investment costs for these processes are 5 to 20 times higher than for biological processes, while the treatment costs are 3 to 10 times higher [24]. In addition, the intermediate products formed during oxidation pose a potential environmental risk [9]. When using AOPs as a pretreatment for the biodegradation of MPs, the advanced oxidation would only serve to alter the surface of the particles, leading to conditions that facilitate the adhesion of microorganisms to the surface (rougher surface and lower hydrophobicity) or transform the polymer into a form that is more susceptible to biodegradation.

In this study, the potentials of two AOPs as pretreatment steps for the biodegradation of PS and PVC MPs by the yeast *Candida parapsilosis* were investigated. The AOPs investigated in this study included treatment with UV-C-activated hydrogen peroxide (UV-C/H_2O_2) and UV-C-activated peroxodisulfate (UV-C/$S_2O_8^{2-}$). These AOPs were selected due to their frequent use in solving problems with recalcitrant organic pollutants and their process characteristics, which are presented below. The efficiencies of the applied pretreatments were evaluated by comparing the intensities of the characteristic bands from the FTIR spectra of the MP samples. The efficiency of the biotreatment step was evaluated by determining the number of colony-forming units.

The UV-C/H_2O_2 process is based on the generation of hydroxyl radicals, HO•, which have a very high standard reduction potential of 2.730 V [25]. The HO• radical is the most studied radical in wastewater treatment, because it is a very reactive, non-selective and ecologically harmless substance with a short lifetime. Furthermore, the rate constants for the reaction between HO• and organic pollutants are in the range of 10^6–10^9 M^{-1} s^{-1} [26], which indicates very rapid oxidation, faster than, for example, when ozone is used for oxidation [27]. During the UV/H_2O_2 process, the photolytic decomposition of each H_2O_2 molecule generates two HO• radicals [28]. H_2O_2 has an absorption maximum at about 254 nm [29,30], which is why UV lamps with a wavelength of 254 nm (UV-C range) are usually used in UV/H_2O_2 processes. In addition to the irradiation wavelength, other factors such as the irradiation intensity, temperature, treatment duration, initial pH of the solution, H_2O_2 dosage, pollutant concentration, matrix, etc., can also influence the efficiency of UV/H_2O_2 treatments [31–34].

The UV/$S_2O_8^{2-}$ process is based on the generation of sulfate radicals, $SO_4•^-$. Although the $SO_4•^-$ radical is a slightly weaker oxidizing agent than the HO• radical, with a standard reduction potential of 2.437 V [25]), the UV/$S_2O_8^{2-}$ process offers several advantages over UV/H_2O_2, making it a worthy candidate for solving the problem of recalcitrant organic pollutants in the environment. $SO_4•^-$ radicals not only have a high oxidizing power, but also react more selectively by electron transfer with organic compounds containing unsaturated bonds or aromatic π electrons. They efficiently degrade organic compounds within a wider pH range. In addition, the $SO_4•^-$ radicals have a half-life of 30–40 µs, which is much longer than 20 ns in the case of HO• radicals [35]. In general, the $SO_4•^-$ radicals can be obtained through the physical or chemical activation of peroxodisulfate ($S_2O_8^{2-}$) or peroxymonosulfate (HSO_5^-) [36]. The photoactivation of $S_2O_8^{2-}$ leads to two $SO_4•^-$ radicals per reactant ion, while in the case of HSO_5^-, the activation generates one $SO_4•^-$ and one HO• radical [37].

The reports on the application of UV/H_2O_2 treatments on PVC MPs are very rare, while the reports on PS MPs are not only rare but also contradictory. Hüffer et al. [38] exposed PS microparticles to a H_2O_2 solution (10% v/v) under 254 nm light for 96 h and reported no significant change in the specific surface area of the exposed particles. However, Dong et al. [39] also investigated the effect of a UV/H_2O_2 treatment on PS MPs and reported significant changes in the physicochemical properties of the sample. In addition, the UV/H_2O_2 treatment resulted in a reduction in the particle size and especially the contact angle of the MPs. This is encouraging with regard to the potential application of this treatment as a pre-step to biotreatment. Hankett et al. [40] investigated the degradation of the plasticizer bis-2-ethylhexyl phthalate (DEHP) in PVC samples using a UV-C/H_2O_2 treatment; the wavelength of irradiation was 254 nm. In addition to the degradation of the plasticizer, they reported radical chain scission reactions with PVC. This scientific team also found that UV-C irradiation not only plays a role in the production of HO• radicals but can also induce molecular changes on the surface of such plasticized PVC samples, leading to increased surface hydrophilicity with an increasing irradiation time [41]. Similar to the UV/H_2O_2 process, there are very few reports on the use of $S_2O_8{}^{2-}$-based AOPs to treat PS and PVC MPs. Liu et al. [42] investigated the effects of thermally activated $S_2O_8{}^{2-}$ treatment on PS microparticles and reported a reduction in the particle size, the formation of holes and cracks on the particle surface and a reduction in the contact angle. Zhang et al. [43] investigated the effect of an $S_2O_8{}^{2-}$-based AOP on the floatability of MPs from various plastic polymers to utilize floatability as a separation method for MP waste. The MPs treated included PVC and PS particles. We did not find any explicit information about the activation method but assume that it was also a thermal activation. The analysis revealed damage to the surface of the plastic particles, followed by the formation of new groups. At optimal conditions of 0.2 M $S_2O_8{}^{2-}$, pH 10, 70 °C and a treatment time of 30 min, the contact angles of PVC and PS particles decreased by 0.13° and 12.20°, respectively, indicating an increased hydrophilicity of the MP surface.

To date, studies on the biodegradation of MPs from conventional plastics have mainly focused on bacteria [16]. More recently, researchers have also turned their attention to other microorganisms, in particular fungi [13]. Similar to bacteria, fungi also produce different types of enzymes [44], which makes them very interesting organisms for testing. Compared to bacteria, however, fungi have interesting peculiarities. For example, they have a much higher metabolic rate [45], and filamentous fungi can secrete specific proteins, so-called hydrophobins, which facilitate the adhesion of their hyphae to hydrophobic surfaces [46]. Regarding the biodegradation of PS MPs by fungi, we found, in one of our previous studies [47], that the yeast *Candida parapsilosis* has the potential to degrade PS and PVC MPs. As far as we know, there are no other reports in the literature on the treatment of PS MPs with yeasts, but only with molds and white-rot fungi [48–55]. Regarding the biodegradation of PVC MPs by fungi, biodegradation by yeasts has not been sufficiently investigated yet; the available studies mostly contain experiments with other fungal species [56–60]. Apart from the report of our working group [51], we did not find any reports on biodegradation by *Candida parapsilosis*. However, we found a report by Webb et al. [61]. They exposed a 0.5 mm thick PVC film to the environment and found that after 95 weeks of exposure, only fungi colonized the PVC. Not a single bacterium was found in the biofilm, indicating that fungi have a greater potential to colonize and biodegrade PVC. Among the fungi isolated from the biofilm, numerous yeasts and yeast-like fungi such as *Rhodotorula aurantiaca, Kluyveromyces* spp. and *Aureobasidium pullulans* were found, pointing to their ability to adapt to PVC MPs.

Although AOP pretreatment prior to biodegradation is not a new concept in environmental engineering, the studies on the contribution of AOP pretreatment to MP biodegradation are very rare and mainly focused on the contribution of plasma treatment or UV irradiation in the presence of atmospheric air [62–65]. Moreover, as far as we know, this is the first such study on MPs from PS or PVC. Considering that the biodegradation of MPs by yeasts has not yet been sufficiently studied, especially when considering the application of the species *Candida parapsilosis*, it is clear that this research represents an exceptional novelty.

2. Materials and Methods

2.1. Reagents and Solutions

The polymers PS and PVC were purchased as the granules DOKI® POLISTIREN 472 (Dioki d.d., Zagreb, Croatia) and GS-28 (Drvoplast d.d., Buzet, Croatia), respectively. A 30% H_2O_2 solution (p.a.; Gram-Mol d.o.o., Zagreb, Croatia) and solid $Na_2S_2O_8$ (p.a.;

Sigma Aldrich, St. Louis, MO, USA) were used as oxidizing agents in the AOP treatments. Also, 0.04 M H_2SO_4, 0.1 M H_2SO_4 and 0.1 M NaOH were used to adjust the pH of the reaction mixture. The H_2SO_4 solutions were prepared from a 98% H_2SO_4 solution (p.a.; Kemika, Zagreb, Croatia), while the NaOH solution was prepared from solid NaOH (p.a.; Sigma Aldrich, St. Louis, MO, USA). A 2% sodium dodecyl sulfate solution and 70% ethanol solution intended for washing the MP samples after biotreatment were prepared from sodium dodecyl sulfate salt (≥99.0%; Sigma Aldrich, St. Louis, MO, USA) and 96% ethanol (Ph. Eur.; Gram-Mol, Zagreb, Croatia), respectively. Sterile ultrapure water (18.2 MΩ cm; Milli-Q™ system with UV lamp, Millipore, Burlington, MA, USA) was used for the experiments.

2.2. Preparation and Characterization of Microplastics

The purchased PS and PVC granules were ground in a cryomill (CryoMill, Retsch, Haan, Germany) with liquid nitrogen at −196 °C and an oscillation frequency of 25 Hz. The ground particles were dried at room temperature (25.0 ± 0.2 °C) for 48 h and then sieved through stainless steel sieves (AS 200 jet, Retsch, Haan, Germany) to obtain MPs in the size range of 25–100 μm. The MPs were then stored in glass bottles until used in the experiments.

An FTIR spectrometer (FTIR-8400S, Shimadzu, Kyoto, Japan) with ATR sampling (MIRacle™ Single Reflection ATR, PIKE Technologies, Fitchburg, WI, USA) was used to characterize the MP particles before and after the treatments.

Prior to the biodegradation experiments, MP particles were sterilized for 10 min in a 100 mL flask containing 70% ethanol; the flask was shaken on a rotary shaker (Incubator 1000 equipped with the platform shaker Unimax 1010, Heidolph Instruments, Schwabach, Germany) at 160 rpm. The sterilized particles were separated from the ethanol using vacuum membrane filtration through a sterile 0.45 μm cellulose nitrate filter (ReliaDisc™ membrane filter, Ahlstrom-Munksjö, Helsinki, Finland).

2.3. AOP Treatments

Two homogeneous AOPs, UV/H_2O_2 and $UV/S_2O_8^{2-}$, were tested for the degradation of PS and PVC MPs. The experimental design used for the AOP experiments is shown in Table 1. In the experiments, the influences of three process parameters were tested: the treatment duration, the initial pH value of the reaction mixture and the initial concentration of peroxide. The experiments were designed according to the full factorial methodology, with each variable (i.e., the process parameter) tested at three levels. The mass of MPs was 55.0 mg in all AOP experiments.

Table 1. Experimental design for $UV\text{-}C/H_2O_2$ and $UV\text{-}C/S_2O_8^{2-}$; the design combines values for the duration of the treatment (t), the initial pH value of the reaction mixture and the initial concentration of peroxide (c).

No.	t/min	pH	c/mM	No.	t/min	pH	c/mM	No.	t/min	pH	c/mM
1	30	4.0	1.0	10	60	4.0	1.0	19	90	4.0	1.0
2	30	4.0	10.5	11	60	4.0	10.5	20	90	4.0	10.5
3	30	4.0	20.0	12	60	4.0	20.0	21	90	4.0	20.0
4	30	7.0	1.0	13	60	7.0	1.0	22	90	7.0	1.0
5	30	7.0	10.5	14	60	7.0	10.5	23	90	7.0	10.5
6	30	7.0	20.0	15	60	7.0	20.0	24	90	7.0	20.0
7	30	10.0	1.0	16	60	10.0	1.0	25	90	10.0	1.0
8	30	10.0	10.5	17	60	10.0	10.5	26	90	10.0	10.5
9	30	10.0	20.0	18	60	10.0	20.0	27	90	10.0	20.0

The experimental procedure was similar for both AOP treatments. First, 55 mg of an MP sample and 80 mL of water were added to the reactor. The pH was then adjusted to the desired value with 0.04 M H_2SO_4, 0.1 M H_2SO_4 or 0.1 M NaOH. The treatment began with the addition of an appropriate amount of peroxide to the reactor to reach the initial concentration specified in the experimental design (Table 1). The peroxide added was H_2O_2 or $Na_2S_2O_8$, depending on the treatment. During the treatment, the reaction mixture was irradiated at 254 nm (UV-C range) with a mercury lamp (Pen-Ray® 11SC-1, UVP, Upland, CA, USA). The homogeneity of the mixture was maintained by mixing at 150 rpm with a high-speed magnetic stirrer (MS-3000, Biosan, Riga, Latvia). The duration of treatment was varied according to the experimental design (Table 1). The treated samples were separated from the reaction mixture using vacuum filtration through a 0.45 µm cellulose nitrate membrane filter (ReliaDisc™, Ahlstrom-Munksjö, Helsinki, Finland), washed with water and dried in air for 24 h. The samples were then analyzed using FTIR-ATR.

Analysis of Influential Parameters and Process Optimization

In order to analyze the influence of the selected process parameters on the treatment efficiency and to determine the optimal process conditions, we decided to develop and apply an empirical mathematical model that describes the dependency between the process parameters and the treatment efficiency based on the Response Surface Methodology (RSM).

In a first step, the procedure involved the development of four mathematical models of different complexities. These models, represented by Equations (1)–(4), linearly combine different contributions of three tested process parameters.

$$y = a_0 + a_1x_1 + a_2x_2 + a_3x_3 \tag{1}$$

$$y = a_0 + a_1x_1 + a_2x_2 + a_3x_3 + a_4x_1x_2 + a_5x_1x_3 + a_6x_2x_3 \tag{2}$$

$$y = a_0 + a_1x_1 + a_2x_2 + a_3x_3 + a_7{x_1}^2 + a_8{x_2}^2 + a_9{x_3}^2 \tag{3}$$

$$y = a_0 + a_1x_1 + a_2x_2 + a_3x_3 + a_4x_1x_2 + a_5x_1x_3 + a_6x_2x_3 + a_7{x_1}^2 + a_8{x_2}^2 + a_9{x_3}^2 \tag{4}$$

To simplify the following discussion, we refer to these models with the Roman numerals I to IV. The coefficients of the models are labeled a, while the tested process parameters are labeled x. Model I represents a simple combination of the linear contributions of the tested process parameters. Model II and Model III are extended by interaction and quadratic terms, respectively. Model IV is the most complex of the models used, as it contains all the aforementioned contributions. We decided not to use more complex models, because the first and second order polynomials are the most commonly used in practice, and introducing additional elements into the RSM model would require an experimental design with a much higher number of experimental cases.

The relative change in intensity (CI) of a selected characteristic FTIR peak, calculated according to Equation (5), was used as an indicator of the degradation of MPs, i.e., as the dependent variable y in the RSM models (Equations (1)–(4)).

$$CI = \frac{A_0 - A}{A_0} \cdot 100\% \tag{5}$$

A_0 is the area under the characteristic FTIR band from the spectrum of the untreated sample, and A is the area under the same band from the spectrum of the treated sample. The intensities of the characteristic bands at 694.4 cm^{-1} (out-of-plane C-H bending vibration) [66] and 1411.9 cm^{-1} (CH_2 bending) [67] were determined for PS and PVC samples, respectively.

The second step of the procedure was to analyze the developed models and select the best model for each of the tested MPs-AOP combinations. These best models were used to determine the optimal values of the process parameters based on the location of the maximum of the response surfaces.

Model development and a statistical analysis of the RSM models were performed using MATLAB R2010b software (The MathWorks, Inc., Natick, MA, USA). Optimizations of the parameters for MP treatments were performed using an interactive interface (MATLAB

Response Surface Modeling Tool; *rstool* command). All calculations were performed with 95% confidence.

2.4. Biodegradation Experiments

The biodegradation experiments were performed on PS and PVC samples that had previously been treated under optimal AOP conditions.

The yeast *Candida parapsilosis* was selected for the biodegradation experiments. The yeast was isolated from an MP-rich environment and cultivated as described in detail in one of our previous reports [47]. It is expected that microorganisms inhabiting an MP-rich environment are already adapted to MPs and that they are more efficient in biodegrading plastic polymers than microorganisms that avoid such environments. The same report [47] shows that *Candida parapsilosis* has the highest potential for biodegrading PS and PVC MPs compared to other yeast cultures isolated from the same environment. The biodegradation experiments were carried out under conditions described as optimal in the same report. These included an agitation speed of 156 rpm and an initial pH of 5.7 for experiments with PS samples, while an agitation speed of 136 rpm and an initial pH of 4.9 were used for experiments with PVC samples. The initial value of the number of colony-forming units per mL (CFU) was in the order of 10^7 in all experiments (corresponding to an optical density value of 1.0).

Each biodegradation experiment was performed in an Erlenmeyer flask (200 mL) that was placed in a thermostatic rotary shaker (Incubator 1000 with Unimax 1010 platform shaker, Heidolph Instruments, Schwabach, Germany) at 25.0 ± 0.2 °C for 30 days. The flask contained a mineral medium, the suspension of yeast *Candida parapsilosis* with an optical density of 1.0 and 40 mg of a pretreated MP sample, giving a total working volume of 80 mL; the initial concentration of MPs in the tested system was therefore 500 mg L^{-1}. The mineral medium was prepared according to Gong et al. [68] and contained 1.000 g L^{-1} K_2HPO_4, 1.000 g L^{-1} KH_2PO_4, 1.000 g L^{-1} NH_4NO_3, 0.500 g L^{-1} NaCl, 0.200 g L^{-1} $MgSO_4{\cdot}7H_2O$, 0.020 g L^{-1} $CaCl_2$ and 0.005 g L^{-1} $FeCl_3{\cdot}6H_2O$. A control flask with the mineral medium, the yeast suspension, but without MPs was prepared for a blank experiment. All experiments were performed in duplicate. At the end of each experiment, the treated MP samples were separated from the aqueous phase using vacuum membrane filtration (0.45 µm filter, ReliaDisc^{TM}, Ahlstrom-Munksjö, Helsinki, Finland) and washed in three washing steps to remove the biomass from the surface of the MPs. The steps included washing in 2% sodium dodecyl sulfate solution, 70% ethanol and sterile water, respectively. Each washing step lasted 30 min and was performed in a 100 mL flask on a rotary shaker (Incubator 1000 with Unimax 1010 platform shaker, Heidolph Instruments, Schwabach, Germany) at 160 rpm.

The CFU value was determined on days 0, 3, 7, 14, 21 and 30 to monitor the growth of *Candida parapsilosis*. The CFU values were determined using the decimal plate method [69].

3. Results and Discussion

To characterize the untreated MP samples, an FTIR-ATR analysis of the samples was performed. Figure S1 shows the recorded FTIR spectra, with distinct bands characteristic for the polymers PS and PVC. In the case of PVC, a strong band at 1720.6 cm^{-1} is observed, indicating an additive in the polymer mass, as the strong band in the range of 1700–1725 cm^{-1} corresponds to the C=O stretching of the carboxylic acids [70], while the PVC structure does not contain such groups.

To evaluate the degradation of MPs, we had to select a characteristic FTIR band whose intensity could be monitored. The selection was made among the three most intense bands from the spectra of the untreated PS and PVC samples (Figure S1). These three bands were the C-H bending bands at 694.4 cm^{-1} and 748.4 cm^{-1} and the aromatic C=C stretching band at 1405.5 cm^{-1} in the PS spectrum (Figure S1, case A), while in the PVC spectrum (Figure S1, case B), they were the C-Cl stretching band at 871.8 cm^{-1}, the C-H bending band at 1280.8 cm^{-1} and the CH_2 bending band at 1411.9 cm^{-1}. A general decrease in the intensities of all three bands from the PS spectra was observed during the AOP treatments. A correlation analysis (Table 2) revealed a high positive correlation between these three PS bands, confirming the existence of similar trends in the decrease of band intensities and showing us that fairly similar results are to be expected regardless of which of these three bands are selected. However, to minimize the influence of random error, we selected the band at 694.4 cm^{-1} to monitor PS degradation, because it has the highest initial intensity. For the PVC samples, the correlation analysis revealed a high positive correlation between the intensities of CH_2 bending and C-Cl stretching, while the behavior of the C-H bending band differed from the other two. It should be noted that a general decrease in the intensities of CH_2 bending and C-Cl stretching was observed in both AOP treatments, while the intensity of the C-H band did not seem to be significantly affected by the treatments. Finally, we followed the same logic as for the PS samples and selected the band with the highest initial intensity, where a decrease in the intensity was evident. In the case of the PVC samples, this was the CH_2 band at 1411.9 cm^{-1}.

Table 2. The results of the correlation analysis of the intensities of the three most intense FTIR bands during AOP treatments of PS and PVC MPs. The analysis was performed with a significance of $p < 0.050$.

		PS				PVC		
AOP		**694.4 cm^{-1}**	**748.4 cm^{-1}**	**1450.5 cm^{-1}**		**871.8 cm^{-1}**	**1280.8 cm^{-1}**	**1411.9 cm^{-1}**
UV-C/H_2O_2	694.4 cm^{-1}	1.0000			871.8 cm^{-1}	1.0000		
	748.4 cm^{-1}	0.9201	1.0000		1280.8 cm^{-1}	−0.0794	1.0000	
	1405.5 cm^{-1}	0.8352	0.9410	1.0000	1411.9 cm^{-1}	0.9860	−0.0723	1.0000
UV-C/$S_2O_8{}^{2-}$	694.4 cm^{-1}	1.0000			871.8 cm^{-1}	1.0000		
	748.43 cm^{-1}	0.9672	1.0000		1280.8 cm^{-1}	0.4849	1.0000	
	1405.5 cm^{-1}	0.9252	0.9540	1.0000	1411.9 cm^{-1}	0.9372	0.5454	1.0000

The analysis of the applicability of the selected AOPs in the degradation of PS and PVC MPs started with the analysis of the influence of the initial pH of the media, the initial concentration of peroxide and the treatment time on the efficiency of the AOPs.

In the case of the UV-C/H_2O_2 AOP, the influence of the pH of the system reflects, in fact, that the reduction potential of the HO•/H_2O system is pH-dependent. Considering the amount of H_2O_2 added, an increase in the H_2O_2 concentration should increase the formation of HO• radicals and consequently improve the degradation of pollutants. However, an excess of H_2O_2 can decrease the oxidation rate due to scavenging side reactions (Equations (6)–(8)).

$$HO\bullet + H_2O_2 \rightleftarrows HO_2\bullet + H_2O \quad (6)$$

$$2HO_2\bullet \rightleftarrows H_2O_2 + O_2 \quad (7)$$

$$HO\bullet + HO_2\bullet \rightleftarrows H_2O + O_2 \quad (8)$$

The perhydroxyl radicals, HO_2•, formed in these reactions have a significantly lower oxidizing power than HO• radicals, since the standard reduction potential of the $HO_2\bullet/H_2O_2$ system is only 1.46 V [25]. In the UV-C/$S_2O_8{}^{2-}$ AOP, the reduction potential of the $SO_4\bullet^-/SO_4{}^{2-}$ system is pH-independent, which means that the pH value should have no influence on the process efficiency. However, the $SO_4\bullet^-$ radical shows a certain degree of autocatalytic behavior, which is most pronounced in the neutral pH range and is suppressed in highly acidic or alkaline solutions [71]. Excessive concentrations of $S_2O_8{}^{2-}$ can lead to a scavenging effect, as shown in Equations (9) and (10) [72].

$$S_2O_8^{2-} + SO_4^{-}\bullet \rightleftarrows SO_2^{2-} + S_2O_8^{-}\bullet \quad (9)$$

$$2SO_4^{-}\bullet \rightleftarrows S_2O_8^{2-} \quad (10)$$

The analysis of the influence of the initial pH of the media, the initial concentration of peroxide and the treatment time required an appropriate mathematical description of the relationship between these process parameters and the relative decrease in intensity of the selected FTIR bands (i.e., the *CI* value). Therefore, a regression diagnosis of four models given by Equations (1)–(4) was first performed. The results of the diagnosis (Table S1) show the significance of all four models, regardless of the MPs and AOP treatments used, as the p-value for each of the models is below the applied significance level of 0.050.

Model I is the simplest of the models examined. The results (Table S1) show that the extension of Model I by interaction terms (Model II) or quadratic terms (Model III) led to a better fit of the *CI* values in most cases. This is reflected by an increase in the values of the coefficient of determination (R^2) or the adjusted coefficient of determination (R^2_{adj}). It is important to note that R^2 is not an ideal indicator of which model is better when comparing models with unequal degrees of freedom. The reason for this is that the R^2 increases with each new independent variable that is added to the model, regardless of whether the added variable is relevant or irrelevant for explaining the variation in the dependent variable [73]. To assess the suitability of the model, we therefore compared the values of R^2_{adj}. R^2_{adj} is the value of the coefficient of determination corrected for the number of predictors used in the model. The protection provided by R^2_{adj} is crucial, as too many terms in a regression model can lead to an "over-fitted" model that provides an incorrect description of the system behavior [73]. The comparison of R^2_{adj} values showed that in three situations (both treatments of the PS samples and the PVC sample treated with UV-C/H_2O_2), the introduction of quadratic terms improved the fit of the *CI* values more than the introduction of interaction terms, indicating the greater importance of the quadratic terms in describing the behavior of these systems. Moreover, the introduction of interaction terms in the case of PS MPs treated with UV-C/$S_2O_8^{2-}$ even decreased the R^2_{adj} value compared to Model I. On the other hand, the contribution of interaction terms proved to be more beneficial in the case of PVC MPs treated with the UV-C/$S_2O_8^{2-}$ process. Finally, Model IV proved to be better than the other three models based on the R^2_{adj} values for all applied AOP treatments, regardless of whether PS or PVC MPs were used. Therefore, this model was used to optimize the AOP treatments.

The estimated values of the coefficients of the RSM models selected as optimal for the applied AOP treatments are presented together with the t-test statistics in Table S2, while the graphical representation of each model's response and the determined optimal values of the process parameters can be seen in Figures 1 and 2.

Since the regression analysis was conducted with a significance of 0.05, and the number of degrees of freedom for the selected RSM models was 17, the corresponding critical value for the two-tailed t-test was 2.110 [74]. The comparison of the t-values for each RSM term with the critical t-value (t-value above the critical value indicates a significant variable) or analysis of the corresponding p-values (p-value below the applied significance of 0.05 indicates a significant variable) shows that all three parameters tested—the treatment duration, concentration of peroxide added and pH of the reaction mixture—have a significant influence on the efficacy of the applied AOP treatments. The duration of the AOP treatment does not show a quadratic relation with the *CI* values in any of the cases examined (all p-values for the a_7 coefficients are above 0.050). In the case of PS MPs, the contribution of the duration is purely linear regardless of the AOP treatment applied; the significance of the duration is only reflected in the coefficient a_1 ($p < 0.050$). Positive values of this coefficient indicate a positive effect of the treatment extension. For PVC treated with UV-C/H_2O_2, the contribution of the duration is reflected by the linear term (coefficient a_1) and in combination with the pH (a_5), whereas for PVC treated with UV-C/$S_2O_8^{2-}$, it is only reflected by the interaction with the concentration of peroxide added (a_4). The coefficients of these interaction terms have a negative sign, which indicates an unfavorable effect of the mutual increase in the values of these process parameters. A linear trend related to the treatment duration, which is visible in the graphical representation of the optimization of the UV-C/$S_2O_8^{2-}$ treatment of PVC MPs (Figure 2, case B2), is obviously not specific to the entire experimental area or statistically insignificant. The influence of the linear contribution of the concentration of peroxide proved to be significant in all four cases investigated ($a_2 < 0.050$). In addition, the significance of the quadratic contribution (a_8) was shown in the case of PVC MPs, regardless of the AOP treatment applied. Negative values of the coefficient a_8 indicate a convex dependence, which is unfavorably influenced by too high or too low of an addition of peroxide. The significance of the quadratic term could not be confirmed for the PS samples. Furthermore, in the PS MPs treated with UV-C/H_2O_2, we observed a significant negative influence of the mutual increase in the concentration of peroxide and the pH of the medium ($a_6 < 0.050$), while in the PVC MPs treated with UV-C/$S_2O_8^{2-}$, a significant negative influence of the mutual increase in the concentration of peroxide and the duration of the treatment was observed ($a_4 < 0.050$). In all four cases examined, the importance of the pH value of the medium is initially reflected in the significance of the linear term (a_3) and the quadratic term (a_9). In the case of the UV-C/H_2O_2 treatment, the importance of pH is also reflected in the interaction terms, which show a combined effect with the concentration of peroxide added in the case of PS MPs (coefficient a_6) or with the treatment duration in the case of PVC MPs (coefficient a_5). The negative sign of coefficient a_9, obtained for all four cases examined, indicates the convex character of the pH influence, which is potentially unfavorable when approaching areas with a higher acidity and basicity.

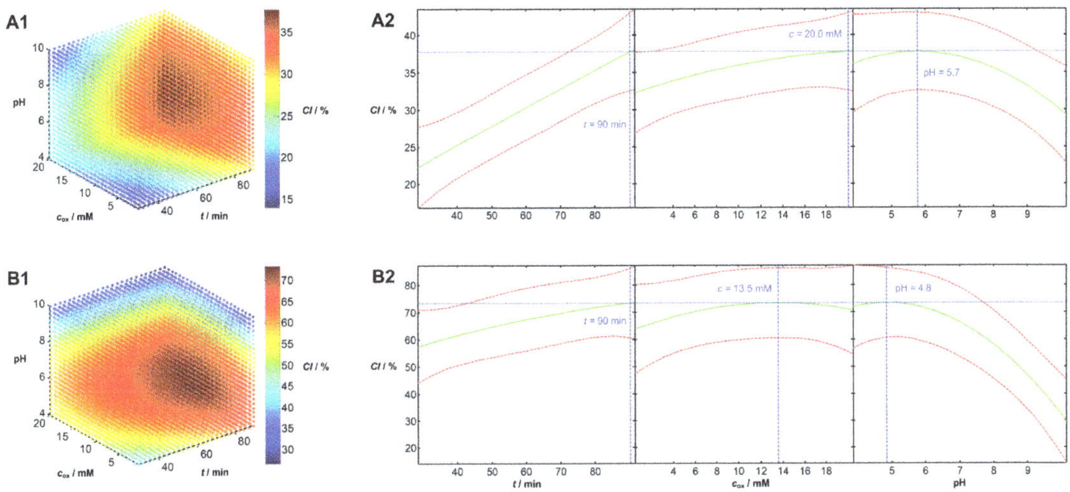

Figure 1. Optimization of UV-C/H_2O_2 treatment of PS MPs (labeled with **A1** and **A2**) and PVC MPs (labeled with **B1** and **B2**). The diagrams under number 1 describe the influence of the process parameters on the treatment efficiency: the response surface is represented with different coloring as the fourth dimension. The diagrams under number 2 show the determination of the optimal process conditions (blue color) by the interactive RSM interface of the MATLAB software: the green line represents the intersection of the response surface, while the red dashed lines are the boundaries of the 95% confidence interval.

The graphical representations of the response surfaces obtained using optimal RSM models are shown in 4D mode (Figures 1 and 2, cases A1 and B1), where the fourth dimension is realized with different coloring: colder colors indicate areas with lower *CI* values, while warmer colors indicate areas with higher *CI* values. The optimal values of the process parameters were determined with 95% confidence using MATLAB's interactive RSM interface and are highlighted in blue in Figures 1 and 2, cases A2 and B2. It can be seen that the experiments with PVC MPs generally resulted in higher *CI* values. Furthermore, the UV-C/H_2O_2 treatment proved to be slightly more effective than the UV-C/$S_2O_8^{2-}$

treatment for the PS samples, while the opposite was true for the PVC samples. As we had previously assumed based on the values of the coefficients of the significant RSM terms, the surface maximum for all four MP-AOP combinations was reached at the longest duration (90 min). The optimum concentration of peroxide was in the high range for PS samples (20.0 mM and 16.8 mM for the UV-C/H_2O_2 treatment and the UV-C/$S_2O_8{}^{2-}$ treatment, respectively) and in the medium range for PVC samples (13.5 mM and 11.1 mM for the UV-C/H_2O_2 treatment and the UV-C/$S_2O_8{}^{2-}$ treatment, respectively). Optimum $S_2O_8{}^{2-}$ concentration values that are below the maximum applied concentration values might result from a scavenging effect according to Equations (9) and (10). The analysis of the influence of the pH showed a favorable effect of a slightly acidic environment for the UV-C/H_2O_2 treatment (optimum pH of 5.7 and 4.8 for PS and PVC samples, respectively) and a neutral to slightly basic environment for the UV-C/$S_2O_8{}^{2-}$ treatment (optimum pH of 7.4 and 8.6 for PS and PVC samples, respectively). The determined optimal pH values in the case of UV-C/$S_2O_8{}^{2-}$ treatment might be related to the autocatalytic activity of $SO_4{\bullet}^-$, which is most pronounced in the neutral pH range [71].

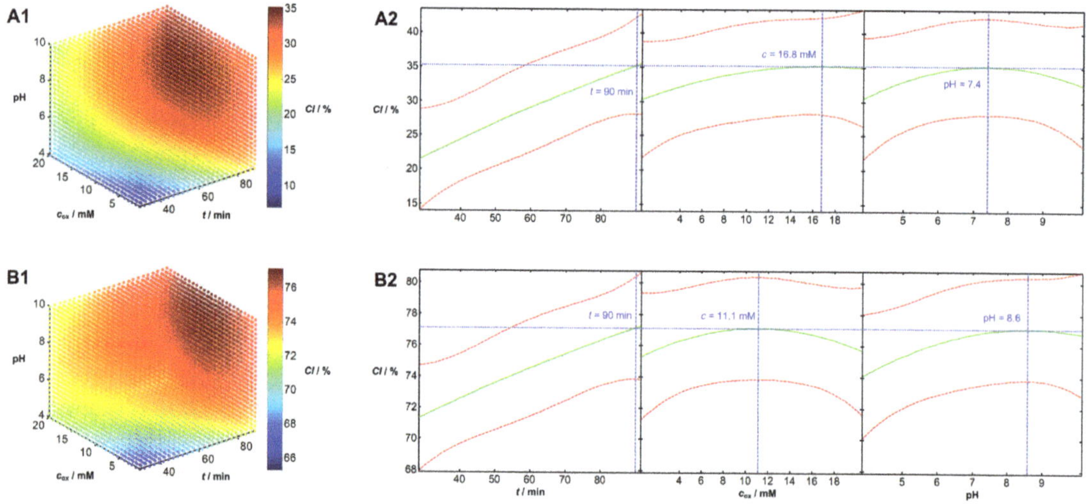

Figure 2. Optimization of UV-C/$S_2O_8{}^{2-}$ treatment of PS MPs (labeled with **A1** and **A2**) and PVC MPs (labeled with **B1** and **B2**). The diagrams under number 1 describe the influence of the process parameters on the treatment efficiency: the response surface is represented with different coloring as the fourth dimension. The diagrams under number 2 show the determination of the optimal process conditions (blue color) by the interactive RSM interface of the MATLAB software: the green line represents the intersection of the response surface, while the red dashed lines are the boundaries of the 95% confidence interval.

In the following phase of this study, *Candida parapsilosis* was used to biologically treat the pretreated MPs samples. These included PS samples subjected to the UV-C/H_2O_2 pretreatment and PVC samples subjected to the UV-C/$S_2O_8{}^{2-}$ pretreatment. The AOP pretreatments were carried out under the optimum conditions determined (Figures 1 and 2, cases A2 and B2). Once the biotreatment of the pretreated PS and PVC samples was completed, the samples were analyzed using FTIR-ATR, and the obtained spectra were compared with the spectra of the untreated and pretreated samples (Figure 3).

The blue line in Figure 3 shows a decrease in the intensity of the monitored FTIR band due to pretreatment, which, as mentioned above, is much more pronounced in the case of PVC. An additional decrease in the intensity of the monitored band (red line) due to biodegradation by *Candida parapsilosis* can also be observed for both PS and PVC samples. However, these findings do not indicate whether the AOP pretreatment contributed to the effectiveness of the biodegradation process. To obtain this information, we compared the CFU values obtained during the biotreatment experiments with pretreated MP samples with the CFU values obtained during the experiments with non-pretreated MP samples (Figure 4). The compared CFU values were corrected for the CFU values of the control experiments. Cases A and B in Figure 4 clearly show that the adaptation of the yeast culture to the conditions in the system took place within the first 3 days and that the number of yeast cells remained practically constant thereafter. An exception is the experiment with the non-pretreated PS sample, where a slight descending trend in the CFU values can be observed, indicating that the yeast culture is unable to adapt to this sample and utilize it as a carbon and energy source. It can also be seen that the CFU levels are generally higher in the pretreated samples, regardless of whether they are PS or PVC samples, suggesting that the changes caused by AOP pretreatment made the MP samples more acceptable for *Candida parapsilosis* survival.

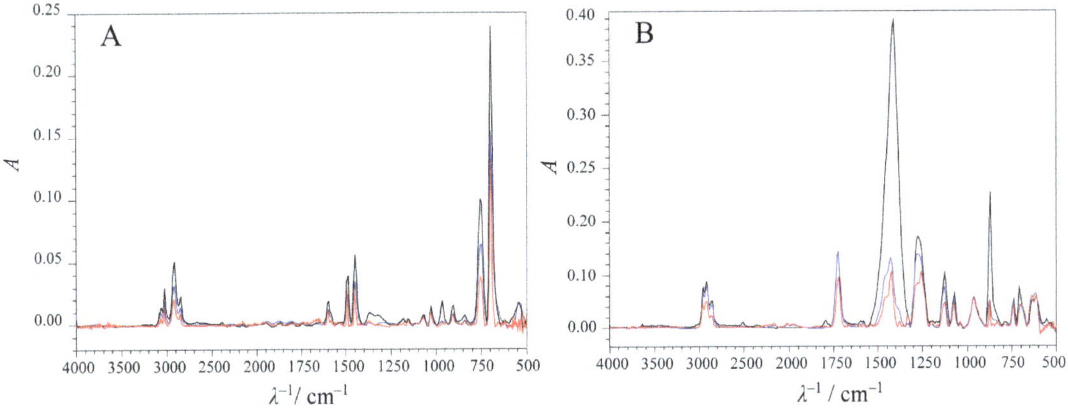

Figure 3. Comparison of FTIR spectra of untreated MP samples (black line), MPs pretreated with selected AOP (blue line) and MPs pretreated with AOP and subsequently biologically treated with *Candida parapsilosis* (red line). The cases presented are (**A**) PS MPs and (**B**) PVC MPs.

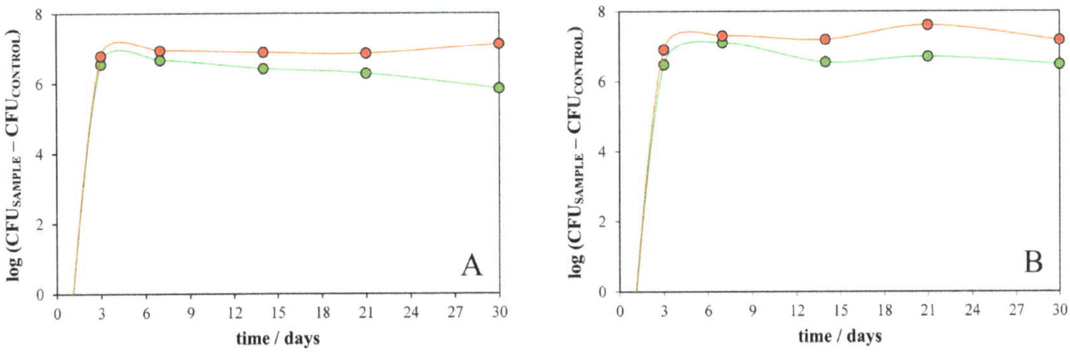

Figure 4. Comparison of CFU values determined during the biodegradation process of non-pretreated (green circles) and AOP-pretreated MP samples (orange circles). The cases shown are (**A**) PS samples and (**B**) PVC samples.

In the end, it seems important to say that *Candida parapsilosis*, the yeast used in this study, is a pathogen and one of the main causes of invasive candidal disease [75]. However, it is a microorganism that is not harmful to humans under normal circumstances, as it is commonly found on the skin, in the gastrointestinal tract and on mucous membranes [76].

The problem can only arise when this microorganism enters a wound. Therefore, the presence of biofilms of *Candida parapsilosis* on the plastic parts of medical equipment is still the main source of infection [77]. The fact that this microorganism can survive on the plastic parts of medical equipment can be seen as an additional argument that *Candida parapsilosis* is able to adapt to MPs.

4. Conclusions

In this study, we tested the applicability of UV-C/H_2O_2 and UV-C/$S_2O_8^{2-}$ AOPs as a pre-step for the biotreatment of PS and PVC MPs by the yeast culture *Candida parapsilosis* isolated from an MP-rich environment. The use of AOPs as a pretreatment implied a restrictive use of AOPs, e.g., to affect only the surface of the MPs, which would not be very costly but could be very beneficial for the efficiency of the biodegradation process. The AOPs mentioned were chosen due to their frequent use in solving a problem with various recalcitrant organic pollutants and due to the high oxidizing power of the radical species involved.

The influence of three process parameters on the efficiency of the applied AOPs in the degradation of MP samples was tested: the initial pH of the solution (tested in the range of 4–10), the initial concentration of peroxide used (1–20 mM) and the treatment duration (up to 90 min). The experimental design followed a full factorial methodology, with each parameter tested at three levels. The efficiency of degradation was evaluated by the relative change in intensity of the selected characteristic FTIR bands: out-of-plane C-H bending at 694.4 cm^{-1} for PS and CH_2 bending at 1411.9 cm^{-1} for PVC samples. Within the applied parameter range, both PS samples and PVC samples were found to be sensitive to the tested photochemical AOPs. A statistical analysis of the treatment results (i.e., the relative change in intensities of selected FTIR bands) confirmed the significant influence of all three tested parameters. The UV-C/H_2O_2 treatment was found to be more effective in degrading PS samples, with a slightly acidic environment, maximum peroxide concentration and maximum treatment duration, leading to the maximum degradation efficiency. For PVC samples, the UV-C/$S_2O_8^{2-}$ treatment led to better treatment results. The maximum efficiency was achieved in a slightly basic environment with a medium peroxide concentration (11.1 mM) for 90 min (maximum treatment duration). Generally speaking, the biodegradation of the pretreated samples resulted in a higher number of yeast cells in the biological treatment system compared to the biological treatment of the non-pretreated samples, indicating a positive effect of the pretreatment on PS and PVC MPs during biodegradation with the yeast *Candida parapsilosis*. The results imply that increasing the duration of the AOP pretreatment or adding more peroxide, as in the case of the UV-C/H_2O_2 treatment of PS MPs, would lead to greater MP degradation during the pretreatment step; this could improve the efficiency of the biological treatment. However, such changes require the delicate consideration of increasing costs (electricity costs, lamp costs and costs of chemicals for oxidation and pH adjustment), and a balance must be found between the increased treatment costs and increased efficiency.

As far as we know, this is the first study on the influence of an AOP pretreatment on the biodegradation of PS and PVC MPs by fungi and one of the few on the biodegradation of AOP-pretreated MPs samples in general. Considering the current societal trends aimed at sustainable development and green technologies, as well as the advantages characterizing treatments such as the one described in the study (high metabolic potential of fungi, very economical and environmentally friendly approach), an intensification of research on the biodegradation of MPs by fungal species is expected. This includes, among other things, a more detailed investigation of various pretreatments that could increase the effectiveness of the biological treatment.

Supplementary Materials: The following supporting information can be downloaded at: https://www.mdpi.com/article/10.3390/w16101389/s1, Figure S1: FTIR spectra of untreated MP samples with the characteristic bands indicated: (A) PS MPs and (B) PVC MPs. The chemical structures of the mentioned polymers are shown in blue; Table S1: Statistical analysis of the fitted response surface models (Equations (1)–(4)) for MPs treated with UV-C/H_2O_2 or UV-C/$S_2O_8^{2-}$. The analysis was performed with a significance of $p < 0.050$; Table S2: Estimated values of the coefficients of the RSM models selected as optimal for the applied AOP treatments (according to Equation (4)) and the results of the *t*-test presented in the form of *p*-values. The calculations were performed with a significance level of $p < 0.050$.

Author Contributions: Conceptualization, M.M. (Marinko Markić), Š.U. and D.K.G.; methodology, K.B.M., M.M. (Martina Miloloža) and M.C.; formal analysis, K.B.M. and V.M.; writing—original draft preparation, K.B.M.; writing—review and editing, Š.U. and D.K.G.; visualization, K.B.M. and Š.U; supervision, T.B. and M.U.B.; project administration, T.B. All authors have read and agreed to the published version of the manuscript.

Funding: The authors would like to acknowledge the financial support of the Croatian Science Foundation through the project Advanced Water Treatment Technologies for Microplastics Removal (AdWaTMiR; IP-2019-04-9661).

Data Availability Statement: Data are contained within the article.

Conflicts of Interest: The authors declare no conflicts of interest.

References

1. Eyerer, P.; Gettwert, V. Properties of Plastics in Structural Components. In *Polymers—Opportunities and Risks I*; Eyerer, P., Ed.; Springer: Berlin/Heidelberg, Germany, 2010; pp. 47–165. [CrossRef]
2. Lamichhane, G.; Acharya, A.; Marahatha, R.; Modi, B.; Paudel, R.; Adhikari, A.; Raut, B.K.; Aryal, S.; Parajuli, N. Microplastics in environment: Global concern, challenges, and controlling measures. *Int. J. Environ. Sci. Technol.* **2023**, *20*, 4673–4694. [CrossRef] [PubMed]
3. Amobonye, A.; Bhagwat, P.; Raveendran, S.; Singh, S.; Pillai, S. Environmental Impacts of Microplastics and Nanoplastics: A Current Overview. *Front. Microbiol.* **2021**, *12*, 768297. [CrossRef] [PubMed]
4. The 17 Goals. Available online: https://sdgs.un.org/goals (accessed on 26 April 2024).
5. Plastic Soup Foundation. Available online: https://www.plasticsoupfoundation.org/en/plastic-problem/sustainable-development/individual-sdgs/ (accessed on 26 April 2024).
6. Alqahtani, S.; Alqahtani, S.; Saquib, Q.; Mohiddin, F. Toxicological impact of microplastics and nanoplastics on humans: Understanding the mechanistic aspect of the interaction. *Front. Toxicol.* **2023**, *5*, 1193386. [CrossRef]
7. Zhao, H.; Hong, X.; Chai, J.; Wan, B.; Zhao, K.; Han, C.; Zhang, W.; Huan, H. Interaction between Microplastics and Pathogens in Subsurface System: What We Know So Far. *Water* **2024**, *16*, 499. [CrossRef]
8. Maddela, N.R.; Kakarla, D.; Venkateswarlu, K.; Megharaj, M. Additives of plastics: Entry into the environment and potential risks to human and ecological health. *J. Environ. Manag.* **2023**, *348*, 119364. [CrossRef] [PubMed]
9. Gao, W.; Zhang, Y.; Mo, A.; Jiang, J.; Liang, Y.; Cao, X.; He, D. Removal of microplastics in water: Technology progress and green strategies. *Green Anal. Chem.* **2022**, *3*, 100042. [CrossRef]
10. Nizzetto, L.; Futter, M.; Langaas, S. Are Agricultural Soils Dumps for Microplastics of Urban Origin? *Environ. Sci. Technol.* **2016**, *50*, 10777–10779. [CrossRef]
11. Iyare, P.U.; Ouki, S.K.; Bond, T. Microplastics removal in wastewater treatment plants: A critical review. *Environ. Sci. Water Res. Technol.* **2020**, *6*, 2664–2675. [CrossRef]
12. Ceretta, M.B.; Nercessian, D.; Wolski, E.A. Current Trends on Role of Biological Treatment in Integrated Treatment Technologies of Textile Wastewater. *Front. Microbiol.* **2021**, *12*, 651025. [CrossRef]
13. Okal, E.J.; Heng, G.; Magige, E.A.; Khan, S.; Wu, S.; Ge, Z.; Zhang, T.; Mortimer, P.E.; Xu, J. Insights into the mechanisms involved in the fungal degradation of plastics. *Ecotoxicol. Environ. Saf.* **2023**, *262*, 115202. [CrossRef]
14. Miri, S.; Saini, R.; Davoodi, S.M.; Pulicharla, R.; Brar, S.K.; Magdouli, S. Biodegradation of microplastics: Better late than never. *Chemosphere* **2022**, *286*, 131670. [CrossRef]
15. Lokesh, P.; Shobika, R.; Omer, S.; Reddy, M.; Saravanan, P.; Rajeshkannan, R.; Saravanan, V.; Venkatkumar, S. Bioremediation of plastics by the help of microbial tool: A way for control of plastic pollution. *Sustain. Chem. Environ.* **2023**, *3*, 100027. [CrossRef]
16. Cai, Z.; Li, M.; Zhu, Z.; Wang, X.; Huang, Y.; Li, T.; Gong, H.; Yan, M. Biological Degradation of Plastics and Microplastics: A Recent Perspective on Associated Mechanisms and Influencing Factors. *Microorganisms* **2023**, *11*, 1661. [CrossRef]
17. Ghosh, S.; Qureshi, A.; Purohit, H.J. Microbial Degradation of Plastics: Biofilms and Degradation Pathways. In *Contaminants in Agriculture and Environment: Health Risks and Remediation*; Kumar, V., Kumar, R., Singh, J., Kumar, P., Eds.; Agro Environ Media: Haridwar, India, 2019; pp. 184–199. [CrossRef]

18. Hanpanich, O.; Wongkongkatep, P.; Pongtharangkul, T.; Wongkongkatep, J. Turning hydrophilic bacteria into biorenewable hydrophobic material with potential antimicrobial activity via interaction with chitosan. *Bioresour. Technol.* **2017**, *230*, 97–102. [CrossRef] [PubMed]
19. Bayry, J.; Aimanianda, V.; Guijarro, J.I.; Sunde, M.; Latgé, J.-P. Hydrophobins—Unique Fungal Proteins. *PLoS Pathog.* **2012**, *8*, e1002700. [CrossRef]
20. Tokiwa, Y.; Calabia, B.P.; Ugwu, C.U.; Aiba, S. Biodegradability of plastics. *Int. J. Mol. Sci.* **2009**, *10*, 3722–3742. [CrossRef]
21. Du, H.; Xie, Y.; Wang, J. Microplastic degradation methods and corresponding degradation mechanism: Research status and future perspectives. *J. Hazard. Mater.* **2021**, *418*, 126377. [CrossRef] [PubMed]
22. Hamd, W.; Daher, E.A.; Tofa, T.S.; Dutta, J. Recent Advances in Photocatalytic Removal of Microplastics: Mechanisms, Kinetic Degradation, and Reactor Design. *Front. Mar. Sci.* **2022**, *9*, 885614. [CrossRef]
23. Mousset, E.; Loh, W.H.; Lim, W.S.; Jarry, L.; Wang, Z.; Lefebvre, O. Cost comparison of advanced oxidation processes for wastewater treatment using accumulated oxygen-equivalent criteria. *Water Res.* **2021**, *200*, 117234. [CrossRef]
24. Rodríguez, M. Fenton and UV-Vis Based Advanced Oxidation Processes in Wastewater Treatment: Degradation, Mineralization and Biodegradability Enhancement. Ph.D. Thesis, University of Barcelona, Faculty of Chemistry, Barcelona, Spain, 2003.
25. Armstrong, D.A.; Huie, R.E.; Koppenol, W.H.; Lymar, S.V.; Merényi, G.; Neta, P.; Ruscic, B.; Stanbury, D.M.; Steenken, S.; Wardman, P. Standard electrode potentials involving radicals in aqueous solution: Inorganic radicals (IUPAC Technical Report). *Pure Appl. Chem.* **2015**, *87*, 1139–1150. [CrossRef]
26. Cheng, M.; Zeng, G.; Huang, D.; Lai, C.; Xu, P.; Zhang, C.; Liu, Y. Hydroxyl radicals based advanced oxidation processes (AOPs) for remediation of soils contaminated with organic compounds: A review. *Chem. Eng. J.* **2016**, *284*, 582–598. [CrossRef]
27. Wang, J.L.; Xu, L.J. Advanced Oxidation Processes for Wastewater Treatment: Formation of Hydroxyl Radical and Application. *Environ. Sci. Technol.* **2012**, *42*, 251–325. [CrossRef]
28. Zoschke, K.; Dietrich, N.; Börnick, H.; Worch, E. UV-based advanced oxidation processes for the treatment of odour compounds: Efficiency and by-product formation. *Water Res.* **2012**, *46*, 5365–5373. [CrossRef] [PubMed]
29. Cataldo, F. Hydrogen peroxide photolysis with different UV light sources including a new UV-LED light source. *New Front. Chem.* **2014**, *23*, 99–110.
30. Aye, T.T.; Low, T.Y.; Sze, S.K. Nanosecond laser-induced photochemical oxidation method for protein surface mapping with mass spectrometry. *Anal Chem.* **2005**, *77*, 5814–5822. [CrossRef] [PubMed]
31. Urbina-Suarez, N.A.; López-Barrera, G.L.; García-Martínez, J.B.; Barajas-Solano, A.F.; Machuca-Martínez, F.; Zuorro, A. Enhanced UV/H_2O_2 System for the Oxidation of Organic Contaminants and Ammonia Transformation from Tannery Effluents. *Processes* **2023**, *11*, 3091. [CrossRef]
32. Račytė, J.; Rimeika, M. Factors influencing efficiency of UV/H_2O_2 advanced oxidation process. In *2007: Proceedings from Kalmar ECO-TECH'07: Technologies for Waste and Wastewater Treatment, Energy from Wast, Remediation of Contaminated Sited, Emissions Related to Climate, Kalmar, Sweden, 26–28 November 2007*; Kaczala, F., Hogland, W., Marques, M., Vinrot, E., Eds.; Linnaeus University: Växjö, Sweden, 2007; pp. 757–769. [CrossRef]
33. Zhou, C.; Gao, N.; Deng, Y.; Chu, W.; Rong, W.; Zhou, S. Factors affecting ultraviolet irradiation/hydrogen peroxide (UV/H_2O_2) degradation of mixed N-nitrosamines in water. *J. Hazard. Mater.* **2012**, *43*, 231–232. [CrossRef] [PubMed]
34. Lopez-Alvarez, B.; Villegas-Guzman, P.; Peñuela, G.A.; Torres-Palma, R.A. Degradation of a Toxic Mixture of the Pesticides Carbofuran and Iprodione by UV/H_2O_2: Evaluation of Parameters and Implications of the Degradation Pathways on the Synergistic Effects. *Water Air Soil Pollut.* **2016**, *227*, 215. [CrossRef]
35. Guerra-Rodríguez, S.; Rodríguez, E.; Singh, D.; Rodríguez-Chueca, J. Assessment of Sulfate Radical-Based Advanced Oxidation Processes for Water and Wastewater Treatment: A Review. *Water* **2018**, *10*, 1828. [CrossRef]
36. Xia, X.; Zhu, F.; Li, J.; Yang, H.; Wei, L.; Li, Q.; Jiang, J.; Zhang, G.; Zhao, Q. A Review Study on Sulfate-Radical-Based Advanced Oxidation Processes for Domestic/Industrial Wastewater Treatment: Degradation, Efficiency, and Mechanism. *Front. Chem.* **2020**, *8*, 592056. [CrossRef]
37. Moreno-Andrés, J.; Rueda-Márquez, J.J.; Homola, T.; Vielma, J.; Moríñigo, M.Á.; Mikola, A.; Sillanpää, M.; Acevedo-Merino, A.; Nebot, E.; Levchuk, I. A comparison of photolytic, photochemical and photocatalytic processes for disinfection of recirculation aquaculture systems (RAS) streams. *Water Res.* **2020**, *181*, 115928. [CrossRef] [PubMed]
38. Hüffer, T.; Weniger, A.-K.; Hofmann, T. Sorption of organic compounds by aged polystyrene microplastic particles. *Environ. Pollut.* **2018**, *236*, 218–225. [CrossRef] [PubMed]
39. Dong, S.; Yan, X.; Yue, Y.; Li, W.; Luo, W.; Wang, Y.; Sun, J.; Li, Y.; Liu, M.; Fan, M. H_2O_2 concentration influenced the photoaging mechanism and kinetics of polystyrene microplastic under UV irradiation: Direct and indirect photolysis. *J. Clean. Prod.* **2022**, *380*, 135046. [CrossRef]
40. Hankett, J.M.; Welle, A.; Lahann, J.; Chen, Z. Evaluating UV/H_2O_2 Exposure as a DEHP Degradation Treatment for Plasticized PVC. *J. Appl. Polym. Sci.* **2014**, *131*, 40649. [CrossRef]
41. Hankett, J.M.; Collin, W.R.; Chen, Z. Molecular Structural Changes of Plasticized PVC after UV Light Exposure. *J. Phys. Chem. B* **2013**, *117*, 16336–16344. [CrossRef] [PubMed]
42. Liu, P.; Qian, L.; Wang, H.; Zhan, X.; Lu, K.; Gu, C.; Gao, S. New insights into the aging behavior of microplastics accelerated by advanced oxidation processes. *Environ. Sci. Technol.* **2019**, *53*, 3579–3588. [CrossRef] [PubMed]

43. Zhang, Y.; Jiang, H.; Wang, H.; Wang, C. Separation of hazardous polyvinyl chloride from waste plastics by flotation assisted with surface modification of ammonium persulfate: Process and mechanism. *J. Hazard. Mater.* **2020**, *389*, 121918. [CrossRef] [PubMed]
44. El-Gendi, H.; Saleh, A.K.; Badierah, R.; Redwan, E.M.; El-Maradny, Y.A.; El-Fakharany, E.M. A Comprehensive Insight into Fungal Enzymes: Structure, Classification, and Their Role in Mankind's Challenges. *J. Fungi* **2021**, *8*, 23. [CrossRef] [PubMed]
45. Gangola, S.; Joshi, S.; Kumar, S.; Pandey, S.C. Chapter 10—Comparative Analysis of Fungal and Bacterial Enzymes in Biodegradation of Xenobiotic Compounds. In *Smart Bioremediation Technologies*; Bhatt, P., Ed.; Academic Press: Chennai, India, 2019; pp. 169–189. [CrossRef]
46. Harding, M.W.; Marques, L.L.R.; Howard, R.J.; Olson, M.E. Can filamentous fungi form biofilms? *Trends Microbiol.* **2009**, *17*, 475–480. [CrossRef] [PubMed]
47. Bule Možar, K.; Miloloža, M.; Martinjak, V.; Cvetnić, M.; Ocelić Bulatović, V.; Mandić, V.; Bafti, A.; Ukić, Š.; Kučić Grgić, D.; Bolanča, T. Bacteria and Yeasts Isolated from the Environment in Biodegradation of PS and PVC Microplastics: Screening and Treatment Optimization. *Environments* **2023**, *10*, 207. [CrossRef]
48. Hock, O.G.; Lum, H.W.; De Qin, D.; Kee, W.K.; Shing, W.L. The growth and laccase activity of edible mushrooms involved in plastics degradation. *Curr. Top. Toxicol.* **2019**, *15*, 57–62.
49. Atiq, N. Biodegradability of Synthetic Plastics Polystyrene and Styrofoam by Fungal Isolates. Ph.D. Thesis, Quaid-i-Azam University, Islamabad, Pakistan, 2011.
50. Chaudhary, A.K.; Vijayakumar, R.P. Studies on biological degradation of polystyrene by pure fungal cultures. *Environ. Dev. Sustain.* **2020**, *22*, 4495–4508. [CrossRef]
51. Chaudhary, A.K.; Chaitanya, K.; Vijayakumar, R.P. Synergistic effect of UV and chemical treatment on biological degradation of Polystyrene by *Cephalosporium* strain NCIM 1251. *Arch. Microbiol.* **2021**, *203*, 2183–2191. [CrossRef]
52. Motta, O.; Proto, A.; De Carlo, F.; De Caro, F.; Santoro, E.; Brunetti, L.; Capunzo, M. Utilization of chemically oxidized polystyrene as co-substrate by filamentous fungi. *Int. J. Hyg. Environ. Health* **2009**, *212*, 61–66. [CrossRef]
53. Galgali, P. Fungal degradation of carbohydrate-linked polystyrenes. *Carbohydr. Polym.* **2004**, *55*, 393–399. [CrossRef]
54. Srikanth, M.; Sandeep, T.S.R.S.; Sucharitha, K.; Godi, S. Biodegradation of plastic polymers by fungi: A brief review. *Bioresour. Bioprocess.* **2022**, *9*, 42. [CrossRef]
55. Ekanayaka, A.H.; Tibpromma, S.; Dai, D.; Xu, R.; Suwannarach, N.; Stephenson, S.L.; Dao, C.; Karunarathna, S.C. A Review of the Fungi That Degrade Plastic. *J. Fungi* **2022**, *8*, 772. [CrossRef]
56. Sakhalkar, S.; Mishra, R. Screening and Identification of Soil Fungi with Potential of Plastic Degrading Ability. *Indian J. Appl. Res.* **2013**, *3*, 62–64. [CrossRef]
57. Malachová, K.; Novotný, Č.; Adamus, G.; Lotti, N.; Rybková, Z.; Soccio, M.; Šlosarčíková, P.; Verney, V.; Fava, F. Ability of *Trichoderma hamatum* Isolated from Plastics-Polluted Environments to Attack Petroleum-Based, Synthetic Polymer Films. *Processes* **2020**, *8*, 467. [CrossRef]
58. Sumathi, T.; Viswanath, B.; Sri Lakshmi, A.; SaiGopal, D.V.R. Production of Laccase by *Cochliobolus* sp. Isolated from Plastic Dumped Soils and Their Ability to Degrade Low Molecular Weight PVC. *Biochem. Res. Int.* **2016**, *2016*, 1–10. [CrossRef] [PubMed]
59. Kirbaş, Z.; Keskin, N.; Güner, A. Biodegradation of polyvinylchloride (PVC) by white rot fungi. *Bull. Environ. Contam. Toxicol.* **1999**, *63*, 335–342. [CrossRef]
60. Ali, M.I.; Ahmed, S.; Robson, G.; Javed, I.; Ali, N.; Atiq, N.; Hameed, A. Isolation and molecular characterization of polyvinyl chloride (PVC) plastic degrading fungal isolates. *J. Basic Microbiol.* **2013**, *54*, 18–27. [CrossRef]
61. Webb, J.S.; Nixon, M.; Eastwood, I.M.; Greenhalgh, M.; Robson, G.D.; Handley, P.S. Fungal Colonization and Biodeterioration of Plasticized Polyvinyl Chloride. *Appl. Environ. Microbiol.* **2000**, *66*, 3194–3200. [CrossRef]
62. Gómez-Méndez, L.D.; Moreno-Bayona, D.A.; Poutou-Piñales, R.A.; Salcedo-Reyes, J.C.; Pedroza-Rodríguez, A.M.; Vargas, A.; Bogoya, J.M. Biodeterioration of plasma pretreated LDPE sheets by *Pleurotus ostreatus*. *PLoS ONE* **2018**, *13*, e0203786. [CrossRef]
63. Scally, L.; Gulan, M.; Weigang, L.; Cullen, P.J.; Milosavljevic, V. Significance of a Non-Thermal Plasma Treatment on LDPE Biodegradation with *Pseudomonas aeruginosa*. *Materials* **2018**, *11*, 1925. [CrossRef]
64. Jeyakumar, D.; Chirsteen, J.; Mukesh, D. Synergistic effects of pretreatment and blending on fungi mediated biodegradation of polypropylenes. *Bioresour. Technol.* **2013**, *148*, 78–85. [CrossRef]
65. Tribedi, P.; Dey, S. Pre-oxidation of low-density polyethylene (LDPE) by ultraviolet light (UV) promotes enhanced degradation of LDPE in soil. *Environ. Monit. Assess.* **2017**, *189*, 624. [CrossRef]
66. Fang, J.; Xuan, Y.; Li, Q. Preparation of polystyrene spheres in different particle sizes and assembly of the PS colloidal crystals. *Sci. China Technol. Sci.* **2010**, *53*, 3088–3093. [CrossRef]
67. Amar, Z.H.; Chabira, S.F.; Sebaa, M.; Ahmed, B. Structural changes undergone during thermal aging and/or processing of unstabilized, dry-blend and rigid PVC, investigated by FTIR-ATR and curve fitting. *Ann. Chim. Sci. Mater.* **2019**, *43*, 59–68. [CrossRef]
68. Gong, Y.; Ding, P.; Xu, M.-J.; Zhang, C.-M.; Xing, K.; Qin, S. Biodegradation of phenol by a halotolerant versatile yeast *Candida tropicalis* SDP-1 in wastewater and soil under high salinity conditions. *J. Environ. Manag.* **2021**, *289*, 112525. [CrossRef] [PubMed]
69. Black, J.G. *Microbiology: Principles and Explorations*, 8th ed.; John Wiley & Sons: Hoboken, NJ, USA, 2012; p. 975.
70. Liu, X. *Organic Chemistry I*; Kwantlen Polytechnic University: Surrey, BC, Canada, 2021; p. 198.
71. Herrera-Ordonez, J. The role of sulfate radicals and pH in the decomposition of persulfate in aqueous medium: A step towards prediction. *Chem. Eng. J. Adv.* **2022**, *11*, 100331. [CrossRef]

72. Honarmandrad, Z.; Sun, X.; Wang, Z.; Naushad, M.; Boczkaj, G. Activated persulfate and peroxymonosulfate based advanced oxidation processes (AOPs) for antibiotics degradation—A review. *Water Resour. Ind.* **2023**, *29*, 100194. [CrossRef]
73. Mahbobi, M.; Tiemann, T.K. Chapter 8. Regression Basics. In *Introductory Business Statistics with Interactive Spreadsheets—1st Canadian Edition*, 1st ed.; Mahbobi, M., Tiemann, T.K., Eds.; Pressbooks: Montreal, QC, Canada; Available online: https://opentextbc.ca/introductorybusinessstatistics/chapter/regression-basics-2/ (accessed on 28 February 2024).
74. Frost, J. Statistics By Jim. T-Distribution Table of Critical Values. Available online: https://statisticsbyjim.com/hypothesis-testing/t-distribution-table/ (accessed on 28 February 2024).
75. Trofa, D.; Gácser, A.; Nosanchuk, J.D. *Candida parapsilosis*, an emerging fungal pathogen. *Clin. Microbiol. Rev.* **2008**, *21*, 606–625. [CrossRef]
76. Branco, J.; Miranda, I.M.; Rodrigues, A.G. *Candida parapsilosis* Virulence and Antifungal Resistance Mechanisms: A Comprehensive Review of Key Determinants. *J. Fungi* **2023**, *9*, 80. [CrossRef]
77. Tóth, R.; Nosek, J.; Mora-Montes, H.M.; Gabaldon, T.; Bliss, J.M.; Nosanchuk, J.D.; Turner, S.A.; Butler, G.; Vágvölgyi, C.; Gácser, A. *Candida parapsilosis*: From Genes to the Bedside. *Clin. Microbiol. Rev.* **2019**, *32*, e00111-18. [CrossRef]

water

MDPI

Article

Evaluation of Fenton, Photo-Fenton and Fenton-like Processes in Degradation of PE, PP, and PVC Microplastics

Kristina Bule Možar [1], Martina Miloloža [1], Viktorija Martinjak [1], Floren Radovanović-Perić [1], Arijeta Bafti [1], Magdalena Ujević Bošnjak [2], Marinko Markić [1], Tomislav Bolanča [1], Matija Cvetnić [1], Dajana Kučić Grgić [1,*] and Šime Ukić [1]

[1] Faculty of Chemical Engineering and Technology, University of Zagreb, Trg Marka Marulića 19, 10000 Zagreb, Croatia; kbule@fkit.unizg.hr (K.B.M.); miloloza@fkit.unizg.hr (M.M.); vprevaric@fkit.unizg.hr (V.M.); fradovano@fkit.unizg.hr (F.R.-P.); abafti@fkit.unizg.hr (A.B.); mmarkic@fkit.unizg.hr (M.M.); tbolanca@fkit.unizg.hr (T.B.); mcvetnic@fkit.hr (M.C.); sukic@fkit.unizg.hr (Š.U.)

[2] Croatian Institute of Public Health, Rockefellerova 7, 10000 Zagreb, Croatia

* Correspondence: dkucic@fkit.unizg.hr; Tel.: +385-14597238

Abstract: The global problem of microplastics in the environment is "inspiring" scientists to find environmentally friendly and economically viable methods to remove these pollutants from the environment. Advanced oxidation processes are among the most promising methods. In this work, the potential of Fenton, photo-Fenton, and Fenton-like processes for the degradation of microplastics from low-density polyethylene (LDPE), polypropylene (PP), and poly(vinyl chloride) (PVC) in water suspensions was investigated. The influence of three parameters on the efficiency of the degradation process was tested: the pH of the medium (3–7), the mass of added iron (10–50 times less than the mass of microplastics), and the mass of added H_2O_2 (5–25 times more than the mass of added iron). The effectiveness of the treatment was monitored by FTIR-ATR spectroscopy. After 60-min treatments, the PP microparticles were found to be insensitive. In the Fenton treatment of PVC and the photo-Fenton treatment of LDPE and PVC, changes in the FTIR spectra related to the degradation of the microplastics were observed. In these three cases, the treatment parameters were optimized. It was found that a low pH (3) and a high iron mass (optimal values were 1/12 and 1/10 of the mass of the microplastics for LDPE and PVC, respectively) favored all three. The degradation of LDPE by the photo-Fenton treatment was favored by high H_2O_2 concentrations (25 times higher than the mass of iron), while these concentrations were significantly lower for PVC (11 and 15 times for the Fenton and photo-Fenton treatment, respectively), suggesting that scavenging activity occurs.

Keywords: microplastics; low-density polyethylene; polypropylene; poly(vinyl chloride); advanced oxidation; Fenton-based processes

1. Introduction

The global development of human society has undoubtedly increased the quality of human life, which is mainly reflected in the prolongation of human life [1,2] and the increase in the world population [3]. On the other hand, this development has led to excessive environmental pollution [4,5], which is a threat to all living organisms, including human beings. One anthropogenic pollutant that has recently become the focus of scientific interest is plastic, especially plastic particles smaller than 5 mm, known as microplastic (MP) particles or simply microplastics (MPs) [6,7]. MPs have so far been detected in various parts of the environment, including in the air [8], soil [9], and water [10]. Because of the high bioavailability of MPs in aquatic medium [11], contamination of this environment is of particular interest to scientists. MP particles can be easily ingested by aquatic organisms and, thus, enter the food chain. Particles bellow 150 μm can be absorbed by biota tissue, organs, and even cells [12]. This leads to a bioaccumulation effect. While it is suspected that there should also be a biomagnification effect, the studies conducted to date have not been able to confirm this [11,13,14]. The adverse potential of MPs has been demonstrated in numerous studies [12,15–17], so it is not surprising that MPs are considered an emerging contaminant [18].

In recent decades, scientists have conducted intensive research into methods that effectively remove MPs from the environment. Of the approaches tested, biological treatment is considered the most environmentally friendly. However, due to the extreme stability of plastics and the hydrophobicity of their surface, it is difficult to find organisms that degrade plastics quickly and efficiently [19]. Another approach is to use membrane processes. However, it has been shown that the removal of MPs using membrane technology is still incomplete [20], and there is also a problem with membrane maintenance and sludge disposal. Recently, the applicability of advanced oxidation processes (AOPs) has been increasingly investigated as an additional option for treating water contaminated with MPs. The basis of AOP treatments is the formation of extremely reactive radicals (strong oxidizing agents), which ensure rapid reaction rates in the degradation of recalcitrant organic pollutants. This approach has a low selectivity and can, therefore, be used for the simultaneous degradation of different pollutants [21]. AOP treatment can lead to partial degradation or, ideally, complete mineralization of the pollutants. Partial degradation usually results in the formation of less complex chemical forms that are more susceptible to other treatment approaches, especially biodegradation, which is the most environmentally friendly approach. For this reason, AOPs are often used in combination with biodegradation as a pretreatment step [22,23].

The AOP studies available in the literature generally do not report complete mineralization of the MPs and focus the discussion on the degree of polymer degradation achieved. Ortiz et al. [24] investigated the effect of Fenton treatment on various MPs samples (polyethylene (PE), polypropylene (PP), poly(vinyl chloride) (PVC), polyethylene terephthalate (PET), and expanded polystyrene (EPS)) in a size range of 150–250 μm obtained from commercially available plastic products. The initial mass of the MPs was 100 mg, the iron concentration was 10 mg/L, and the H_2O_2 concentration was 1000 mg/L. The treatment lasted 7.5 h. During the treatment, additional amounts of iron and H_2O_2 were added to the system at regular intervals to recover the consumed reagents. After the treatment under environmental conditions (pH and temperature) showed no changes in the samples, they decided to perform the treatment under more extreme conditions: an acidic medium with a pH of 3 and elevated temperature (80 °C). This treatment resulted in a mass loss of about 8 to 12%. Lang et al. [25] investigated the effect of Fenton treatment (7 days, room temperature, pH 4) on the adsorption of heavy metals on the surface of PS MPs and reported that the applied treatment contributes to the aging of MPs. Piazza et al. [26] studied the environmental aspects of a photo-Fenton treatment of PP and PVC MPs (the samples had a size of 155 and 73 μm, respectively). The treatment involved visible light irradiation and the application of ZnO nanorods coated with a SnO_x layer and decorated with Fe^0 nanoparticles. The authors monitored the changes in the FTIR spectra to determine the percentage of degradation of the MPs. After 7 days of treatment, a degradation efficiency of 94–96% was observed. Miao et al. [27] reported a 56% mass loss of PVC microparticles with a size of 100–200 μm and a 75% dechlorination after 6 h treatment with an electro-Fenton-like system based on a TiO_2/graphite cathode. They also observed a significant positive contribution of temperature increase to the degree of dechlorination. In addition, the literature also reports on the investigation of the applicability of numerous other AOP treatments [28–35].

In this study, the potentials of three Fenton-based AOPs, namely Fenton, photo-Fenton, and Fenton-like processes were evaluated for the degradation of MP particles of low-density PE (LDPE), PP, and PVC. The influence of the pH of the medium, the quantity of added oxidizing agents, and the MPs concentration on the degradation process was investigated, and the optimum treatment conditions were determined.

2. Fenton-Based Processes

AOPs based on the Fenton reaction are widely used for the oxidation of organic macro- and micropollutants [36]. Some of the main advantages of Fenton-based processes are the relatively simple process equipment, relatively inexpensive and environmentally friendly reagents, and a relatively simple treatment procedure. In addition, these processes are conducted at room temperature and atmospheric pressure, which provides an additional economic advantage. Fenton-based processes are known to be highly resistant to matrix interferences and typically result in high mineralization rates for recalcitrant organic pollutants [37–39].

The basis of Fenton-based processes is the generation of the highly reactive hydroxyl radical (•OH) with a standard redox potential of $E^0(\bullet OH/H_2O)$ = 2.730 V [40]. The oxidation activity of the hydroxyl radical is related to the pH of the solution (Equation (1)), so that the redox potential of the system increases with decreasing pH.

$$\bullet OH + H^+ + e^- \rightarrow H_2O \quad (1)$$

In the classical Fenton process (Figure 1, case A), the Fenton reagent is responsible for the formation of hydroxyl radicals. The Fenton reagent is a mixture of aqueous solutions of an iron(II) salt and H_2O_2, in which hydroxyl radicals are formed according to Equation (2) [41].

$$Fe^{2+} + H_2O_2 + H^+ \rightarrow Fe^{3+} + \bullet OH + H_2O \quad (2)$$

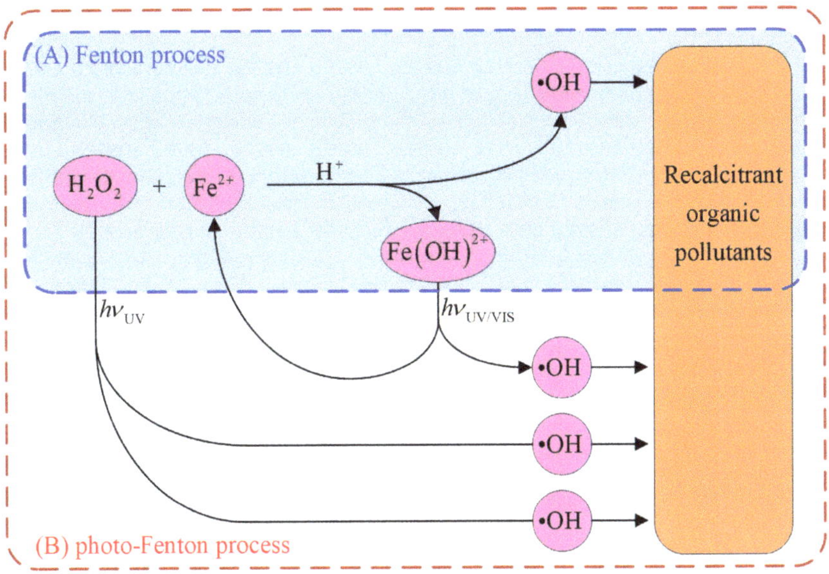

Figure 1. Schematic representation of (**A**) the Fenton process (outlined in blue) and (**B**) the photo-Fenton process (outlined in red).

The formation of hydroxyl radicals can be negatively affected by an excess of H_2O_2 or Fe^{2+} ions in the bulk (Equations (3) and (4)) [37,39], causing an undesirable scavenging effects, like the one described by Equation (5) [42].

$$\bullet OH + H_2O_2 \rightarrow HO_2\bullet + H_2O \quad (3)$$

$$\bullet OH + Fe^{2+} + H^+ \rightarrow Fe^{3+} + H_2O \quad (4)$$

$$HO_2\bullet + \bullet OH \rightarrow H_2O_2 + O_2 \quad (5)$$

An alkaline medium is not suitable for carrying out the Fenton process because it reduces not only the oxidizing power of the hydroxyl radical but also the activity of the Fenton reagent. More specifically, at a high pH, the concentration of Fe^{2+} ions in the solution is negatively affected by the formation of iron(II) hydroxide precipitate. In addition, high pH values favor the decomposition of H_2O_2. However, very low pH values are also not beneficial for the activity of the Fenton reagent. This is because at very low pH values, the active H_2O_2 concentration is affected by the formation of stable $H_3O_2^+$ ions [43]. Therefore, in order to avoid or minimize undesirable scavenging effects and to increase the efficiency of the treatment, the process conditions must be optimized. Due to the specificity of the system to be treated, optimization should be performed for each system.

The inclusion of ultraviolet or visible light irradiation in the classical Fenton process leads to a photo-Fenton process (Figure 1, case B) in which hydroxyl radicals are increasingly produced due to a two-step reaction. First, the iron(III) ions from Equation (2) are hydrolyzed, leading to the formation of the $Fe(OH)^{2+}$ complex (Equation (6)), which subsequently undergoes photo-induced reduction and generates hydroxyl radical (Equation (7)) [44,45].

$$Fe^{3+} + H_2O \rightarrow Fe(OH)^{2+} + H^+ \quad (6)$$

$$Fe(OH)^{2+} + h\nu \rightarrow Fe^{2+} + \bullet OH \quad (7)$$

In addition, UV irradiation can cause photolysis of H_2O_2, which also increases the production of hydroxyl radicals (Equation (8)) [43].

$$H_2O_2 + h\nu \rightarrow 2(\bullet OH) \quad (8)$$

Despite the advantage of the increased production of hydroxyl radicals and the corresponding expected higher efficiency in the degradation of organic pollutants, the photo-Fenton process has some disadvantages. First and foremost, it requires irradiation during treatment, which results in high energy consumption. In addition, high concentrations of organic pollutants in a treated system can significantly reduce the absorption of irradiation by the $Fe(OH)^{2+}$ complex, resulting in an increase in irradiation time. All this makes photo-Fenton treatment much more expensive than classical Fenton treatment.

In addition to the photo-Fenton process, there are numerous other modifications of the classical Fenton process, commonly referred to as Fenton-like processes. Fenton-like processes are divided into heterogeneous and homogeneous processes. In heterogeneous Fenton-like processes, the Fe^{2+} ion in the Fenton reagent is replaced by a solid catalyst, while homogeneous Fenton-like processes refer to processes in which a different metal ion is used in place of the iron ion in combination with H_2O_2 [46]. Among heterogeneous Fenton-like systems, systems triggered by zero-valent iron, especially in the form of nanoparticles, are very popular, mainly because of their large specific surface area and high reactivity [43,47]. The process starts with the reaction between zero-valent iron and H_2O_2, leading to the formation of Fe^{2+} ions (Equation (9)), which are crucial for the formation of the hydroxyl radical (Equation (2)).

$$Fe^0 + H_2O_2 + 2H^+ \rightarrow Fe^{2+} + 2H_2O \quad (9)$$

The Fe^{3+} ions formed in the reaction described by Equation (2) additionally contribute to the formation of the Fe^{2+} ions (Equation (10)) [43].

$$Fe^0 + 2Fe^{3+} \rightarrow 3Fe^{2+} \quad (10)$$

It should be noted, however, that there is concern in the scientific community about the potential adverse effects that the presence of nanoparticles of zero-valent iron may cause in the environment [48,49].

3. Materials and Methods

3.1. Reagents and Solutions

The plastic material was purchased in the form of granules as OKITEN® 245 A (Dioki d.d., Zagreb, Croatia), GF10 (Xiamen Keyuan Plastic Co., Ltd., Xiamen, China), and GS-28 (Drvoplast d.d., Buzet, Croatia) for LDPE, PP, and PVC, respectively. Ferrous sulfate heptahydrate ($FeSO_4 \cdot 7H_2O$; ≥99%; Sigma-Aldrich, Burlington, MA, USA), elemental iron (≥95%; Carlo Erba Reagents, Milan, Italy), 30% H_2O_2 solution (1.11 g cm^{-3}; Gram-Mol d.o.o., Zagreb, Croatia), 0.1 M NaOH solution (Lach-Ner s.r.o., Naratovice, Czech Republic), and H_2SO_4 solution (Kemika, Zagreb, Croatia) at concentrations of 0.1 and 5 M were used for the AOP treatments. Ultrapure water (18.2 MΩ cm; Milli-Q, Millipore, Burlington, MA, USA) was used for all experiments.

3.2. Preparation and Characterisation of Microplastics

The purchased plastic granules were crushed in a cryo-mill (Retsch, Haan, Germany) with liquid nitrogen at an operating temperature of −196 °C. After grinding, the MPs were dried at room temperature (25.0 ± 0.2 °C) for 48 h before sieving. The ground plastic was sieved on stainless steel sieves (AS 200 jet, Retsch, Hann, Germany) to obtain MPs in the size range of 25–100 μm, which were then used for the experiments.

FTIR-ATR analysis (FTIR-8400S, Shimadzu, Kyoto, Japan, and MIRacle™ Single Reflection ATR, PIKE Technologies, Madison, WI, USA) was used to characterize the untreated MP particles and the MP particles after AOP treatments.

To gain additional insight into the MPs' degradation, the untreated MP particles and the MP particles treated under optimal conditions were analyzed by scanning electron microscopy (SEM) using a Tescan Vega Easyprobe 3 microscope (Brno, Czech Republic). Imaging was performed in secondary electron mode (SE) and backscattered electron mode (BSE) at an accelerating voltage of 10 kV and a working distance of about 8 mm. Prior to imaging, Pd/Au was sputtered onto the samples using a Quorum Technologies SC7620 sputter coater (Lewes, UK) at 18 mA for 60 s.

3.3. AOP Treatments

The experimental design for the AOP treatments (Table 1) followed the full factorial methodology with three variables tested at three levels, resulting in a total of 27 trials for each AOP treatment applied. The variables tested were pH, mass ratio of MPs to Fe, and mass ratio of H_2O_2 to Fe. Although it would be appropriate to express the amounts of iron and peroxide in the reaction system as a molar concentration (due to the fact that the potential of the system depends on the molar concentration of the substance), the AOP experiments were performed in such a way that we could only estimate the molar concentrations of these substances. Although it is known that the volume of the reaction mixture does not deviate too much from 80 mL, i.e., from the amount of added water, we cannot know the true volume of the reaction mixture because we also add a certain quantity of MPs, Fe, and H_2O_2 to the system and adjust the pH of the reaction mixtures at the beginning of the experiment. Therefore, we decided to express the amounts of the mentioned substances in the reaction system in terms of their mass instead of their molar concentration. The range of tested values (i.e., levels) was selected based on the information available in the literature [50,51]. The required masses of iron species and H_2O_2 volumes were calculated based on the mass of MPs, which was 55 mg for all experiments.

The experimental procedure was similar for all three AOP treatments. The predefined mass of 55 mg of an MPs sample was added to the reactor together with 80 mL of water. The pH was adjusted to the desired value with 0.1 M H_2SO_4, 5 M H_2SO_4, or 0.1 M NaOH, after which a certain amount of iron in a suitable form was also added to the reactor. In the Fenton and photo-Fenton treatments, ferrous ion (Fe^{2+}) was used, which was added in the form of $FeSO_4 \cdot 7H_2O$. For the Fenton-like process, elemental iron (Fe^0) was used. In the final step, an appropriate amount of 30% H_2O_2 was added to initiate the treatment. In the photo-Fenton treatment only, the reaction mixture was exposed to UV-C irradiation of 254 nm (mercury lamp Pen-Ray® 11SC-1, UVP, Upland, CA, USA). All AOP experiments lasted 60 min, with continuous mixing at 150 rpm using a magnetic stirrer (high speed magnetic stirrer MS-3000, Biosan SIA, Riga, Latvia). After each experiment, the MPs were separated from the aqueous phase by vacuum membrane filtration through a sterile 0.45 μm cellulose nitrate membrane filter (ReliaDisc™ membrane filter, Ahlstrom, Helsinki, Finland). The MPs were then washed with water and air-dried for 24 h prior to FTIR-ATR analysis.

Table 1. Experimental design for Fenton, photo-Fenton, and Fenton-like processes.

No.	pH	m(MPs):m(Fe)	$m(H_2O_2)$:m(Fe)	No.	pH	m(MPs):m(Fe)	$m(H_2O_2)$:m(Fe)	No.	pH	m(MPs):m(Fe)	$m(H_2O_2)$:m(Fe)
1	3.0	10:1	5:1	10	4.5	10:1	5:1	19	6.0	10:1	5:1
2	3.0	10:1	15:1	11	4.5	10:1	15:1	20	6.0	10:1	15:1
3	3.0	10:1	25:1	12	4.5	10:1	25:1	21	6.0	10:1	25:1
4	3.0	25:1	5:1	13	4.5	25:1	5:1	22	6.0	25:1	5:1
5	3.0	25:1	15:1	14	4.5	25:1	15:1	23	6.0	25:1	15:1
6	3.0	25:1	25:1	15	4.5	25:1	25:1	24	6.0	25:1	25:1
7	3.0	50:1	5:1	16	4.5	50:1	5:1	25	6.0	50:1	5:1
8	3.0	50:1	15:1	17	4.5	50:1	15:1	26	6.0	50:1	15:1
9	3.0	50:1	25:1	18	4.5	50:1	25:1	27	6.0	50:1	25:1

3.4. Determination of Optimal Conditions

The optimal conditions for the degradation of LDPE, PP, and PVC MPs by the three applied Fenton-based AOPs were determined using the response surface modeling (RSM) approach. The RSM approach involves a series of mathematical techniques aimed at building an empirical model that describes the relationship between the input variables (i.e., the independent variables) and one or more dependent variables (i.e., the response) [52]. Based on our experience in the field of optimization of pollutant degradation in the aquatic environment [53,54], we have assumed that a quadratic model represented by Equation (11) should be sufficient to describe the system with three independent variables and one dependent variable.

$$y = a_0 + a_1x_1 + a_2x_2 + a_3x_3 + a_4x_1x_2 + a_5x_1x_3 + a_6x_2x_3 + a_7{x_1}^2 + a_8{x_2}^2 + a_9{x_3}^2 \quad (11)$$

The model coefficients are denoted by a, while x_1 to x_3 represent the values of the three process parameters tested: pH, mass ratio of MPs to iron, and mass ratio of H_2O_2 to iron. In addition to the linear (x_1, x_2, x_3) and quadratic (${x_1}^2$, ${x_2}^2$, ${x_3}^2$) contributions, the model also includes the interaction terms (x_1x_2, x_1x_3, x_2x_3) to cover the possible combined effects of the parameters tested, which, as we knew from experience [53–57], are not uncommon in such systems. The area of a broad FTIR band that occurred in the spectral range 3000–3600 cm^{-1} was used as the response, i.e., as the dependent variable y.

Statistical analysis of developed model was performed using Design-Expert 10.0 software (StatEase, Minneapolis, MN, USA).

4. Results and Discussion

First, the untreated MP samples were characterized by FTIR-ATR spectroscopy. The recorded spectra (Figure 2) clearly show the absorption bands characteristic of the polymers LDPE, PP, and PVC. The spectrum of untreated PVC (Figure 2C) contains an unexpected peak in the spectral range of 1715–1750 cm^{-1}, which corresponds to the stretching of the carbonyl group [58]. This peak obviously represents an additive present in the PVC sample tested, as pure PVC has no carbonyl groups in its structure.

Each treated sample was also analyzed by FTIR-ATR to detect changes in the intensity of the bands related to the vibrations of the groups formed during the degradation of the MPs. The formation of carbonyl species is one of the most important indicators of the oxidative degradation of MPs, and the quantification of the formed carbonyl groups by infrared spectroscopy has been the method of choice for many years due to the strong absorption in the spectral range of 1715–1750 cm^{-1}, which is related to the stretching of the carbonyl group [58–61]. However, since the untreated PVC samples showed a peak in this region (Figure 2C), we decided not to use this band as an indicator of MPs' degradation, but rather a broad band that appeared in the spectral range 3000–3600 cm^{-1}, with a maximum at about 3370 cm^{-1}. The appearance of the band in this region indicates side reactions, such as the substitution of chloride by hydroxide [33,62] and the formation of the terminal C≡C–H group [33]. According to Liu [58], the band of OH stretching is expected in the range of 3200–3600 cm^{-1}, but the width and position of the band strongly depend on the amount of hydrogen bonding [63]. We have found several reports on the occurrence of this band during the degradation of plastics [33,62,64,65]. The stretching of the terminal C≡C–H group is expected in the range of 3250–3350 cm^{-1} [58] and may, therefore, be hidden in a broad band of OH stretching.

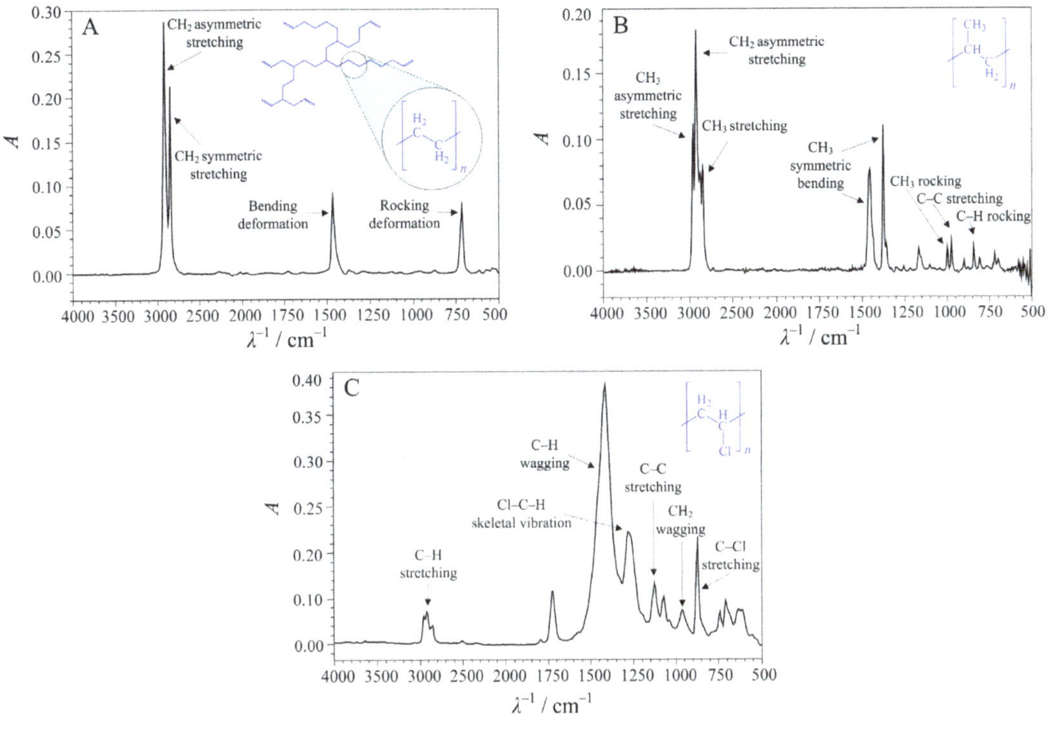

Figure 2. FTIR spectra of untreated MPs with characteristic peaks of (**A**) LDPE, (**B**) PP, and (**C**) PVC. The molecular structures of the polymers are shown in blue in the top righthand corner of each spectrum.

After analyzing the obtained FTIR spectra, it was found that some of the applied treatments, especially the Fenton-like treatment, did not result in significant spectral changes that can be associated with the degradation of MPs (Table 2). This is not so surprising, since the MPs studied are composed of polymers that do not have hydrolysable groups in their structure and are, therefore, less susceptible to degradation [66]. PP samples remained unaffected by all three treatments applied. For LDPE, only the photo-Fenton treatment led to observable changes in the intensity of the monitored band, which is apparently due to the introduction of UV-C irradiation in the treatment. Apart from the fact that the introduction of UV-C irradiation increases the production of radicals compared to the conventional Fenton process, UV-C irradiation alone may also have a significant impact on the degradation of MPs [67]. Of the three MPs tested, PVC was found to be the most susceptible to Fenton-based treatments, with changes in the intensity of the monitored band observed for the Fenton and photo-Fenton treatments. The reason for this is most likely the chlorine atom contained in the PVC structure, which facilitates the oxidation and decomposition of PVC [66]. According to the available information, the first step of PVC decomposition should be dechlorination or dehydrochlorination, during which various chlorinated compounds can be formed [27,68]. Furthermore, dehydrochlorination leads to the formation of labile internal allyl chloride structures [69], which dissociate into chlorine radicals, Cl•, and polyene radicals, R• (Equation (12)).

$$R - Cl \rightarrow R\bullet + Cl\bullet \quad (12)$$

Polyene radicals react with oxygen from the environment and form peroxy radicals (ROO•), which can react with –CH_2– or –CH–Cl groups and, thus, contribute to the further degradation of PVC [68].

Table 2. Comparison of the experiments performed with regard to the changes observed in the FTIR bands.

Polymer	Treatment		
	Fenton	Photo-Fenton	Fenton-like
LDPE	NO	YES	NO
PP	NO	NO	NO
PVC	YES	YES	NO

Considering what is discussed above, we report here only the RSM models developed for the treatment cases that resulted in observable changes in the intensity of the monitored band. Statistical analysis of these models (Tables 3–5) confirmed their significance for describing the influence of the three variables tested. The adequacy of the models is further confirmed by the fact that 87.06 to 93.40% of the variance in the response (i.e., the independent variable), according to the R^2 values, can be explained by the variability of the values of the tested (i.e., dependent) variables. The negative values of the coefficient a_1 for all three models clearly indicate a positive influence of a more acidic environment on the efficiency of the applied treatments. The negative values of coefficient a_2 obtained for the treatments of PVC samples indicate the positive influence of a higher concentration of Fe^{2+} ions in the system, while the negative values of coefficient a_3 indicate an unfavorable influence of increased H_2O_2 concentrations, suggesting the potential scavenging effect of H_2O_2 during PVC treatment. No such effects were observed in the case of LDPE treatment.

Figures 3–5 contain graphical representations of the response surfaces corresponding to the models whose parameters are listed in Tables 3–5, while the optimal conditions for each treatment, estimated from the maxima of the response surfaces, are listed in Table 6. Visualizing the response surface of a system that has four dimensions (three independent and one dependent variable) is a challenge. Therefore, for the sake of simplicity, we decided to visualize the response surfaces at fixed values of one of the independent variables, namely pH. In this way, we enabled a transparent analysis of the behavior of the response by plotting three surfaces. In the first place, the surface trends can be seen based on the variation in the parameters m(MPs):m(Fe) and $m(H_2O_2)$:m(Fe). Secondly, by comparing three response surfaces, it is possible to see what happens to the response due to the variation in pH. The area, A, of a broad FTIR band in the spectral range 3000–3600 cm^{-1} represents the response. The red areas on the surfaces show the most favorable conditions for performing a Fenton or photo-Fenton treatment, while the blue areas represent the least favorable conditions. The response surfaces confirm some of our conclusions derived from the statistical analysis of the RSM models. For example, the most intense red areas are obtained at a pH of 3. Furthermore, the response surfaces show that a high amount of added iron (small ratio m(MPs):$m(Fe^{2+})$) is beneficial not only for PVC treatment but also for

LDPE treatment. This is in line with some other reports claiming that a high concentration of Fe^{2+} ions is required for the degradation of organic pollutants by the homogeneous Fenton process [26,70]. Finally, in the case of LDPE treated with the photo-Fenton process (Figure 3), a positive effect of increased H_2O_2 concentrations is observed, with the optimal amount of added H_2O_2 being 25 times higher than the amount of added Fe (Table 6). In the case of PVC treatment, the red areas cover low to medium H_2O_2 concentrations, regardless of whether we analyze the Fenton or the photo-Fenton response surface. The optimal amounts of added H_2O_2 are estimated to be 11 and 15 times the amount of added Fe for the Fenton and photo-Fenton treatments, respectively.

Table 3. Statistical analysis of the fitted response surface model (Equation (11)) for the case of LDPE MPs treated with the photo-Fenton process. The analysis was performed with a significance of $p < 0.050$.

Model				**Coefficients**		**Influential Variables** [1]		
R^2	R^2_{adj}	F	p	**Value**	p	x_1	x_2	x_3
0.8706	0.8021	12.71	<0.0001	$a_0 = 4.51$	-			
				$a_1 = -2.63$	<0.0001	YES		
				$a_2 = 0.55$	0.1534			
				$a_3 = 1.14$	0.0061			YES
				$a_4 = 2.52$	<0.0001	YES	YES	
				$a_5 = -0.30$	0.5109			
				$a_6 = 0.13$	0.7692			
				$a_7 = 2.51$	0.0010	YES		
				$a_8 = -1.02$	0.1247			
				$a_9 = -0.28$	0.6640			

Note: [1] x_1 = pH; $x_2 = m(MPs){:}m(Fe)$; $x_3 = m(H_2O_2){:}m(Fe)$.

Table 4. Statistical analysis of the fitted response surface model (Equation (11)) for the case of PVC MPs treated with the Fenton process. The analysis was performed with a significance of $p < 0.050$.

Model				**Coefficients**		**Influential Variables** [1]		
R^2	R^2_{adj}	F	p	**Value**	p	x_1	x_2	x_3
0.8994	0.8462	16.89	<0.0001	$a_0 = 18.42$	-			
				$a_1 = -4.64$	<0.0001	YES		
				$a_2 = -7.12$	<0.0001		YES	
				$a_3 = -1.76$	0.0289			YES
				$a_4 = 0.44$	0.6347			
				$a_5 = 0.64$	0.4879			
				$a_6 = 0.86$	0.3531			
				$a_7 = -1.70$	0.2009			
				$a_8 = 0.35$	0.7878			
				$a_9 = -3.90$	0.0071			YES

Note: [1] x_1 = pH; $x_2 = m(MPs){:}m(Fe)$; $x_3 = m(H_2O_2){:}m(Fe)$.

Table 5. Statistical analysis of the fitted response surface model (Equation (11)) for the case of PVC MPs treated with the photo-Fenton process. The analysis was performed with a significance of $p < 0.050$.

Model				Coefficients		Influential Variables [1]		
R^2	R^2_{adj}	F	p	Value	p	x_1	x_2	x_3
0.9340	0.8990	26.72	<0.0001	$a_0 = 22.29$	-			
				$a_1 = -5.35$	<0.0001	YES		
				$a_2 = -7.00$	<0.0001		YES	
				$a_3 = -0.76$	0.2249			
				$a_4 = -0.92$	0.2265			
				$a_5 = -0.43$	0.5969			
				$a_6 = 1.06$	0.1690			
				$a_7 = -1.80$	0.1022			
				$a_8 = -0.11$	0.9185			
				$a_9 = -4.25$	0.0008			YES

Note: [1] x_1 = pH; x_2 = m(MPs):m(Fe); x_3 = $m(H_2O_2)$:m(Fe).

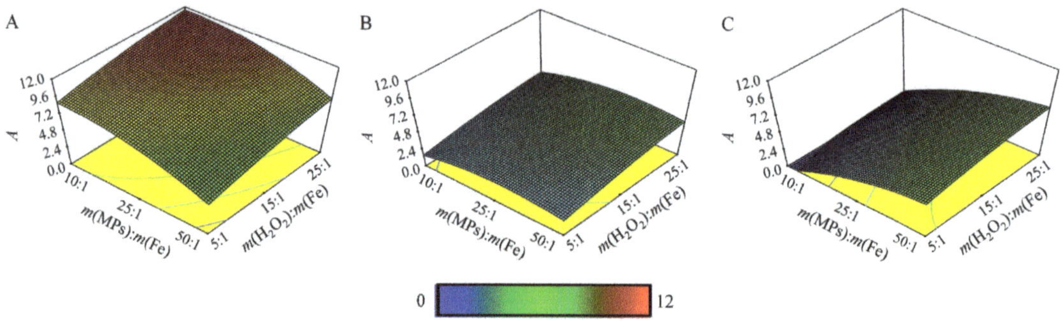

Figure 3. Obtained response surfaces for treatment of LDPE MPs by Fenton process for (**A**) pH = 3.0, (**B**) pH = 4.5, and (**C**) pH = 6.0.

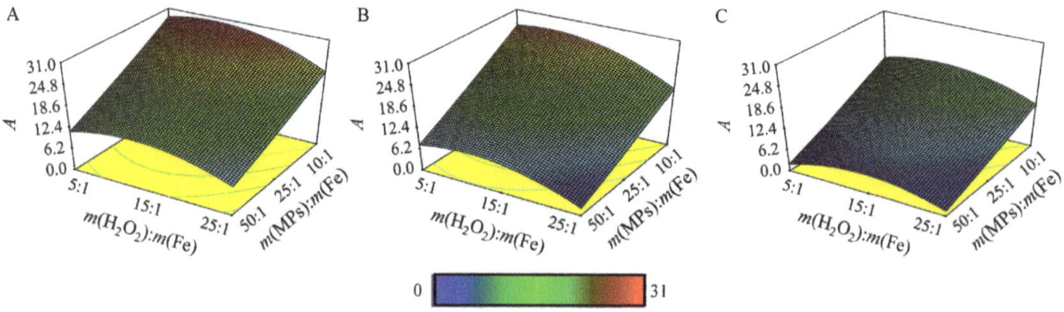

Figure 4. Obtained response surfaces for treatment of PVC MPs by Fenton process for (**A**) pH = 3.0, (**B**) pH = 4.5, and (**C**) pH = 6.0.

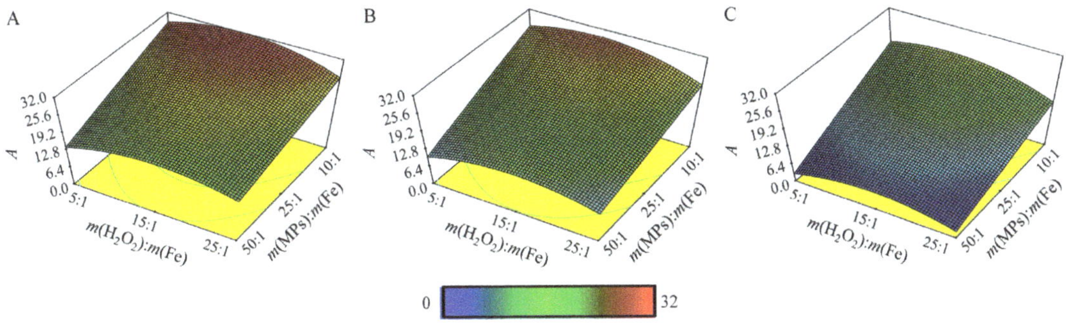

Figure 5. Obtained response surfaces for treatment of PVC MPs by photo-Fenton process for (**A**) pH = 3.0, (**B**) pH = 4.5, and (**C**) pH = 6.0.

Table 6. Optimal conditions for the treatment of MPs samples.

Polymer	Treatment	pH	m(MPs):m(Fe)	$m(H_2O_2)$:m(Fe)
LDPE	Photo-Fenton	3	12:1	25:1
PVC	Fenton	3	10:1	11:1
	Photo-Fenton	3	10:1	15:1

For final confirmation of the results obtained, all three treatments of MPs (LDPE with Fenton and PVC with Fenton and photo-Fenton treatments) were performed under optimal conditions. The FTIR spectra and SEM images of the MPs samples after the treatments were recorded and compared with those of the untreated samples (Figures 6–8).

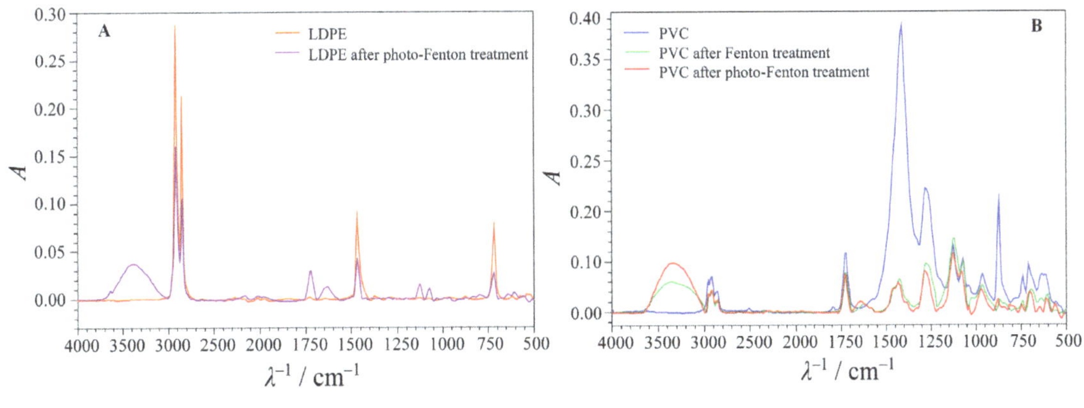

Figure 6. Comparison of FTIR spectra obtained before and after treatment under optimal conditions. The cases show (**A**) LDPE MPs and (**B**) PVC MPs.

When analyzing the FTIR spectra obtained, in addition to a clear increase in the intensity of the band in the spectral range of 3000–3600 cm^{-1} as a result of the treatments, a decrease in the intensity of the characteristic bands of LDPE and PVC polymers can also be seen. Also, a new peak appeared at 1620 cm^{-1} for all three treated samples. The peak in this region is mainly related to C=C stretching, which also indicates polymer degradation [27,71]. In the case of LDPE, the appearance of C=C structures indicates dehydrogenation as one of the degradation reactions; we found that dehydrogenation during PE degradation has also been confirmed in some other treatments [72,73]. The formation of carbonyl groups was confirmed by a new peak at 1715 cm^{-1} (Figure 6A).

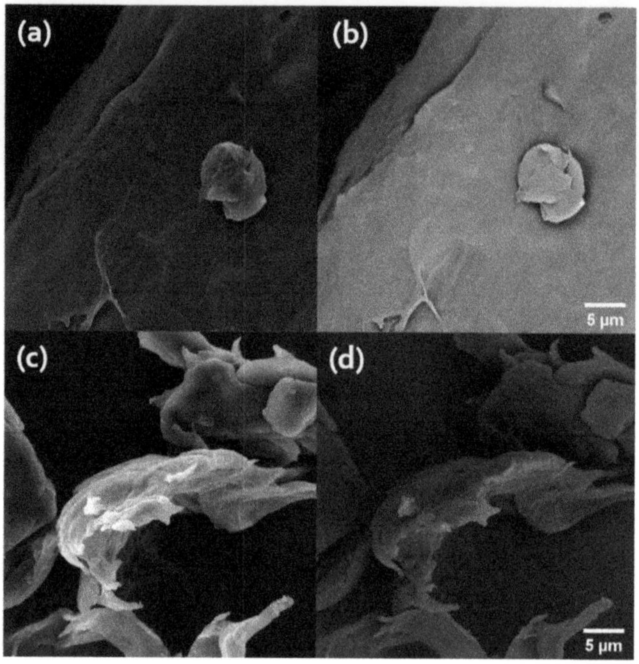

Figure 7. SEM micrographs of LDPE samples—untreated sample scanned in (**a**) SE and (**b**) BSE mode, and photo-Fenton-treated sample scanned in (**c**) SE and (**d**) BSE mode.

The decrease in the intensity of the bands of C–Cl stretching (834 cm^{-1} [74,75]), CH wagging (1427 cm^{-1} [75]), and CH_2 wagging (964 cm^{-1} [75]) in the case of treated PVC supports our assumption that dehydrogenation or dehydrochlorination are the initial steps of PVC degradation. Moreover, this corresponds to the formation of the C=C peak at 1620 cm^{-1}, since dehydrochlorination leads to the formation of polyene structures [68,76]. The comparison of the intensities of the monitored band for the PVC treatments shows the higher efficiency of the photo-Fenton treatment compared to the classical Fenton treatment; the apparent reason for this is the introduction of UV-C radiation into the treatment. PVC is considered to be more sensitive to photo-treatment compared to the other two polymers investigated, especially in the wavelength range of 253–310 nm [68], where the irradiation used in this study is located.

In the SEM analysis, we imaged the MPs samples in SE and BSE mode (Figures 7 and 8). Most of the morphologically and texturally relevant changes on the surface of the polymer samples could be evidenced with the SE detector, while compositionally (chemically) relevant changes could be observed with the BSE detector. Drastic changes in the chemical composition were unlikely, but BSE images could indicate relevant areas of degradation. Textural changes were expected.

As can be seen in Figure 7a, the untreated LDPE microparticles have a low surface roughness, and no agglomerates or aggregates are present. Figure 7b shows that the phase composition was highly homogeneous. The photo-Fenton treatment of the LDPE sample led to an increase in surface roughness with additional morphological defects (Figure 7c). The BSE image (Figure 7d) potentially depicts a gradient, indicating a change in composition compared to the reference.

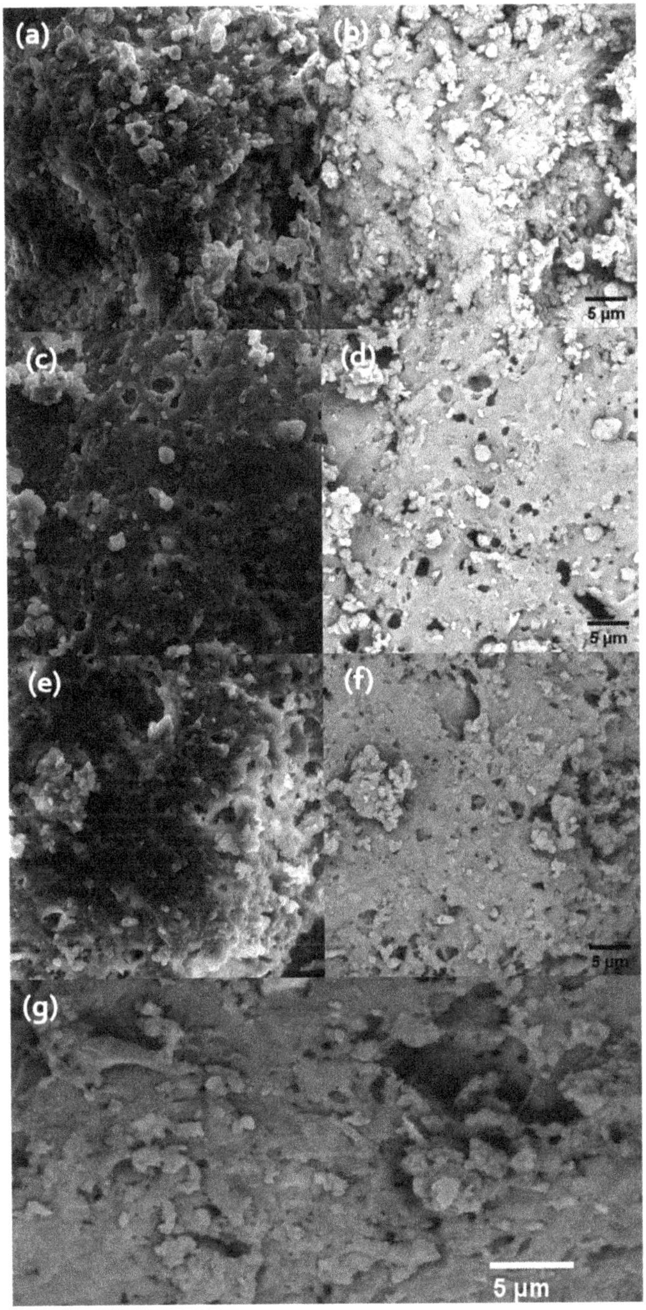

Figure 8. SEM micrographs of PVC samples—untreated sample scanned in (**a**) SE and (**b**) BSE mode, Fenton-treated sample scanned in (**c**) SE and (**d**) BSE mode, and photo-Fenton-treated sample scanned in (**e**) SE and (**f**,**g**) BSE mode.

In contrast to the untreated LDPE MPs, the surface morphology of the untreated PVC MPs scanned in SE mode (Figure 8a) shows considerable roughness and a grainy morphology, while BSE scanning (Figure 8b) reveals a homogeneous phase composition. Moreover, the influences of the Fenton treatment and the photo-Fenton treatment on the MPs were more evident in the PVC samples. In the case of the Fenton treatment, an increase in surface roughness and apparent porosity can be observed (Figure 8c). The BSE images point to the absence of any compositional phase changes (Figure 8d). In the photo-Fenton treatment (Figure 8e,f), however, the porosity increases even more, together with the surface roughness. The BSE image (Figure 8f) shows possible in-depth compositional phase changes. Figure 8g shows an enlarged view of an area with the most relevant morphological changes observed in Figure 8f. The extended degradation is visible, i.e., the porosity extends beyond the surface of the samples.

5. Conclusions

AOPs are an emerging approach for the remediation of environments polluted by non-biodegradable organic substances [77]. Conventional plastic polymers, especially in the form of micro- and nanoplastics, are non-biodegradable organic pollutants whose presence in the environment has become a global environmental problem [78].

In this study, the potential of Fenton, photo-Fenton, and Fenton-like processes for the degradation of LDPE, PP, and PVC MPs was investigated without going into the analysis of the degradation mechanisms. The treatments lasted 60 min. PP was the least sensitive to the treatments, with no significant changes observed in the intensities of the characteristic FTIR bands. For LDPE, signs of degradation were only observed after the photo-Fenton treatment. The PVC samples were the most sensitive to the treatments, as signs of degradation were observed after the Fenton and photo-Fenton treatments, while the Fenton-like treatment showed no effect. The results indicate that of the three treatments applied, the photo-Fenton treatment is the most efficient in degrading PVC MPs. The SEM analysis of the LDPE samples treated with photo-Fenton and the PVC samples treated with Fenton and photo-Fenton confirmed the changes on the MP surface and even the in-depth changes in the PVC treated with photo-Fenton.

The Fenton-based processes did not prove to be overly efficient in degrading LDPE, PP, and PVC MPs at the selected treatment duration of 60 min. It is likely that extending the treatment would lead to better efficiency, but in this case the question of the economic viability of the treatment arises. Therefore, investigating the potential of some other AOPs, e.g., those combining UV-C radiation with other oxidizing agents, such as ozone, peroxide, or persulfate, seems to be a more promising way to remediate the environment polluted by MPs. However, since this study has shown that the Fenton treatment of PVC and the photo-Fenton treatment of LDPE and PVC have led to some changes in MPs, there is a possibility that these treatments can be efficiently used as pretreatments for biodegradation, as the observed changes could be very beneficial for microbial colonization of the surface of MPs.

Author Contributions: Conceptualization, M.M. (Marinko Markić), Š.U., and D.K.G.; methodology, K.B.M., M.M. (Martina Miloloža), and M.C.; formal analysis, K.B.M., F.R.-P., A.B., and V.M.; writing—original draft preparation, K.B.M.; writing—review and editing, Š.U. and D.K.G.; visualization, K.B.M. and Š.U.; supervision, T.B. and M.U.B.; project administration, T.B. All authors have read and agreed to the published version of the manuscript.

Funding: The authors would like to acknowledge the financial support of the Croatian Science Foundation through the Advanced Water Treatment Technologies for Microplastics Removal project (AdWaTMiR; IP-2019-04-9661).

Data Availability Statement: Data are contained within the article.

Conflicts of Interest: The authors declare no conflicts of interest.

References

1. Finch, C.E. Evolution of the human lifespan and diseases of aging: Roles of infection, inflammation, and nutrition. *Proc. Natl. Acad. Sci. USA* **2010**, *107*, 1718–1724. [CrossRef]
2. Blagosklonny, M.V. Why human lifespan is rapidly increasing: Solving "longevity riddle" with "revealed-slow-aging" hypothesis. *Aging* **2010**, *2*, 177–182. [CrossRef] [PubMed]
3. United Nations Population Fund, World Population Trends. Available online: https://www.unfpa.org/world-population-trends#readmore-expand (accessed on 20 February 2023).
4. Ukaogo, P.O.; Ewuzie, U.; Onwuka, C.V. Environmental pollution: Causes, effects, and the remedies. In *Microorganisms for Sustainable Environment and Health*; Chowdhary, P., Raj, A., Verma, D., Akhter, Y., Eds.; Elsevier: Chennai, India, 2020; pp. 419–429. [CrossRef]
5. Martinez, J.L. Environmental pollution by antibiotics and by antibiotic resistance determinants. *Environ. Pollut.* **2009**, *157*, 2893–2902. [CrossRef] [PubMed]
6. Chen, J.; Wu, J.; Sherrell, P.C.; Chen, J.; Wang, H.; Zhang, W.-x.; Yang, J. How to Build a Microplastics-Free Environment: Strategiesfor Microplastics Degradation and Plastics Recycling. *Adv. Sci.* **2022**, *9*, 2103764. [CrossRef] [PubMed]
7. Erceg, M.; Tutman, P.; Vojanić Varezić, D.; Bobanović, A. Karakterizacija mikroplastike u sedimentu plaže Prapratno. *Kem. Ind.* **2020**, *69*, 253–260. [CrossRef]
8. Gasperi, J.; Wright, S.L.; Dris, R.; Collard, F.; Mandin, C.; Guerrouache, M.; Langlois, V.; Kelly, F.J.; Tassin, B. Microplastics in air: Are we breathing it in? *Curr. Opin. Environ. Sci. Health* **2018**, *1*, 1–5. [CrossRef]
9. Boots, B.; Russell, C.W.; Green, D.S. Effects of Microplastics in Soil Ecosystems: Above and Below Ground. *Environ. Sci. Technol.* **2019**, *53*, 11496–11506. [CrossRef] [PubMed]
10. Scherer, C.; Weber, A.; Stock, F.; Vurusic, S.; Egerci, H.; Kochleus, C.; Arendt, N.; Foeldi, C.; Dierkes, G.; Wagner, M.; et al. Comparative assessment of microplastics in water and sediment of a large European river. *Sci. Total Environ.* **2020**, *738*, 139866. [CrossRef]
11. Akhbarizadeh, R.; Moore, F.; Keshavarzi, B. Investigating microplastics bioaccumulation and biomagnification in seafood from the Persian Gulf: A threat to human health? *Food Addit. Contam. A* **2019**, *36*, 1696–1708. [CrossRef]
12. Yuan, Z.; Nag, R.; Cummins, E. Human health concerns regarding microplastics in the aquatic environment—From marine to food systems. *Sci. Total Environ.* **2022**, *823*, 153730. [CrossRef]
13. Miller, M.E.; Hamann, M.; Kroon, F.J. Bioaccumulation and biomagnification of microplastics in marine organisms: A review and meta-analysis of current data. *PLoS ONE* **2020**, *15*, e0240792. [CrossRef] [PubMed]
14. Covernton, G.A.; Cox, K.D.; Fleming, W.L.; Buirs, B.M.; Davies, H.L.; Juanes, F.; Dudas, S.E.; Dower, J.F. Large size (>100 μm) microplastics are not biomagnifying in coastal marine food webs of British Columbia, Canada. *Ecol. Appl.* **2022**, *32*, e2654. [CrossRef] [PubMed]
15. Cui, W.; Gao, P.; Zhang, M.; Wang, L.; Sun, H.; Liu, C. Adverse effects of microplastics on earthworms: A critical review. *Sci. Total Environ.* **2022**, *850*, 158041. [CrossRef]
16. Zolotova, N.; Kosyreva, A.; Dzhalilova, D.; Fokichev, N.; Makarova, O. Harmful effects of the microplastic pollution on animal health: A literature review. *PeerJ* **2022**, *10*, e13503. [CrossRef]
17. Lei, L.; Liu, M.; Song, Y.; Lu, S.; Hu, J.; Cao, C.; Xie, B.; Shi, H.; He, D. Polystyrene (nano)microplastics cause size-dependent neurotoxicity, oxidative damage and other adverse effects in *Caenorhabditis elegans*. *Environ. Sci. Nano* **2018**, *5*, 2009–2020. [CrossRef]
18. Rubio-Armendáriz, C.; Alejandro-Vega, S.; Paz-Montelongo, S.; Gutiérrez-Fernández, Á.J.; Carrascosa-Iruzubieta, C.J.; Hardisson-de la Torre, A. Microplastics as Emerging Food Contaminants: A Challenge for Food Safety. *Int. J. Environ. Res. Public Health* **2022**, *19*, 1174. [CrossRef]
19. Bule Možar, K.; Miloloža, M.; Martinjak, V.; Cvetnić, M.; Kušić, H.; Bolanča, T.; Kučić Grgić, D.; Ukić, Š. Potential of Advanced Oxidation as Pretreatment for Microplastics Biodegradation. *Separations* **2023**, *10*, 132. [CrossRef]
20. Poerio, T.; Piacentini, E.; Mazzei, R. Membrane Processes for Microplastic Removal. *Molecules* **2019**, *24*, 4148. [CrossRef]
21. Lupu, G.-I.; Orbeci, C.; Bobirică, L.; Bobirică, C.; Pascu, L.F. Key Principles of Advanced Oxidation Processes: A Systematic Analysis of Current and Future Perspectives of the Removal of Antibiotics from Wastewater. *Catalysts* **2023**, *13*, 1280. [CrossRef]
22. Volke-Sepúlveda, T.; Saucedo-Castañeda, G.; Gutiérrez-Rojas, M.; Manzur, A.; Favela-Torres, E. Thermally treated low density polyethylene biodegradation by *Penicillium pinophilum* and *Aspergillus niger*. *J. Appl. Polym. Sci.* **2002**, *83*, 305–314. [CrossRef]
23. Esmaeili, A.; Pourbabaee, A.A.; Alikhani, H.A.; Shabani, F.; Esmaeili, E. Biodegradation of low-density polyethylene (LDPE) by mixed culture of Lysinibacillus xylanilyticus and Aspergillus niger in soil. *PLoS ONE* **2013**, *8*, e71720. [CrossRef]
24. Ortiz, D.; Munoz, M.; Nieto-Sandoval, J.; Romera-Castillo, C.; de Pedro, Z.M.; Casas, J.A. Insights into the degradation of microplastics by Fenton oxidation: From surface modification to mineralization. *Chemosphere* **2022**, *309*, 136809. [CrossRef]
25. Lang, M.; Yu, X.; Liu, J.; Xia, T.; Wang, T.; Jia, H.; Guo, X. Fenton aging significantly affects the heavy metal adsorption capacity of polystyrene microplastics. *Sci. Total Environ.* **2020**, *722*, 137762. [CrossRef] [PubMed]
26. Piazza, V.; Uheida, A.; Gambardella, C.; Garaventa, F.; Faimali, M.; Dutta, J. Ecosafety Screening of Photo-Fenton Process for the Degradation of Microplastics in Water. *Front. Mar. Sci.* **2022**, *8*, 791431. [CrossRef]
27. Miao, F.; Liu, Y.; Gao, M.; Yu, X.; Xiao, P.; Wang, M.; Wang, S.; Wang, X. Degradation of polyvinyl chloride microplastics via an electro-Fenton-like system with a TiO_2/graphite cathode. *J. Hazard. Mater.* **2020**, *399*, 123023. [CrossRef] [PubMed]

28. Liu, X.; Sun, P.; Qu, G.; Jing, J.; Zhang, T.; Shi, H.; Zhao, Y. Insight into the characteristics and sorption behaviours of aged polystyrene microplastics through three type of accelerated oxidation processes. *J. Hazard. Mater.* **2021**, *407*, 124836. [CrossRef] [PubMed]
29. Amelia, D.; Karamah, E.F.; Mahardika, M.; Syafri, E.; Rangappa, S.M.; Siengchin, S.; Asrofi, M. Effect of advanced oxidation process for chemical structure changes of polyethylene microplastics. *Mater Today Proc.* **2021**, *52*, 2501–2504. [CrossRef]
30. Wang, Q.; Chen, M.; Min, Y.; Shi, P. Aging of polystyrene microplastics by UV/sodium percarbonate oxidation: Organic release, mechanism, and disinfection by-product formation. *J. Hazard. Mater.* **2024**, *464*, 132934. [CrossRef]
31. de Aragão Belé, T.G.; Neves, T.F.; Cristale, J.; Prediger, P.; Constapel, M.; Dantas, R.F. Oxidation of microplastics by O_3 and O_3/H_2O_2: Surface modification and adsorption capacity. *J. Water Process Eng.* **2021**, *41*, 102072. [CrossRef]
32. Ouyang, Z.; Li, S.; Zhao, M.; Wangmu, Q.; Ding, R.; Xiao, C.; Guo, X. The aging behavior of polyvinyl chloride microplastics promoted by UV-activated persulfate process. *J. Hazard. Mater.* **2022**, *424*, 127461. [CrossRef]
33. Kang, J.; Zhou, L.; Duan, X.; Sun, H.; Ao, Z.; Wang, S. Degradation of cosmetic microplastics via functionalized carbon nanosprings. *Matter* **2019**, *1*, 745–758. [CrossRef]
34. Luo, H.; Zeng, Y.; Zhao, Y.; Xiang, Y.; Li, Y.; Pan, X. Effects of advanced oxidation processes on leachates and properties of microplastics. *J. Hazard. Mater.* **2021**, *413*, 125342. [CrossRef] [PubMed]
35. Zhou, L.; Wang, T.; Qu, G.; Jia, H.; Zhu, L. Probing the aging processes and mechanisms of microplastic under simulated multiple actions generated by discharge plasma. *J. Hazard. Mater.* **2020**, *398*, 122956. [CrossRef] [PubMed]
36. Kastanek, F.; Spacilova, M.; Krystynik, P.; Dlaskova, M.; Solcova, O. Fenton Reaction–Unique but Still Mysterious. *Processes* **2023**, *11*, 432. [CrossRef]
37. Deng, Y.; Zhao, R. Advanced oxidation processes (AOPs) in wastewater treatment. *Curr. Pollut. Rep.* **2015**, *1*, 167–176. [CrossRef]
38. Hamd, W.; Daher, E.A.; Tofa, T.S.; Dutta, J. Recent Advances in Photocatalytic Removal of Microplastics: Mechanisms, Kinetic Degradation, and Reactor Design. *Front. Mar. Sci.* **2022**, *9*, 885614. [CrossRef]
39. Mohammed Bello, M.; Abdul Raman, A.A.; Asghar, A. A review on approaches for addressing the limitations of Fenton oxidation for recalcitrant wastewater treatment. *Process Saf. Environ. Prot.* **2019**, *126*, 119–140. [CrossRef]
40. Armstrong, D.A.; Huie, R.E.; Koppenol, W.H.; Lymar, S.V.; Merényi, G.; Neta, P.; Ruscic, B.; Stanbury, D.M.; Steenken, S.; Wardman, P. Standard electrode potentials involving radicals in aqueous solution: Inorganic radicals (IUPAC Technical Report). *Pure Appl. Chem.* **2015**, *87*, 1139–1150. [CrossRef]
41. Kovacic, M.; Kusic, H.; Loncaric Bozic, A.; Dionysiou, D.D. Advanced Oxidation Processes. In *Encyclopedia of Water: Science, Technology, and Society*; Maurice, P.A., Ed.; Wiley: Hoboken, NJ, USA, 2019; Volume 4, pp. 1925–1940. [CrossRef]
42. Garcia-Costa, A.L.; Casas, J.A. Intensification strategies for thermal H_2O_2-based advanced oxidation processes: Current trends and future perspectives. *Chem. Eng. J. Adv.* **2022**, *9*, 100228. [CrossRef]
43. Xu, M.; Wu, C.; Zhou, Y. Advancements in the Fenton Process for Wastewater Treatment. In *Advanced Oxidation Processes—Applications, Trends, and Prospects*; Bustillo-Lecompte, C., Ed.; IntechOpen: Rijeka, Croatia, 2020; pp. 1–17. [CrossRef]
44. Patel, D.; Wu, J.; Chan, P.; Upreti, S.; Turcotte, G.; Ye, T. Surface modification of low density polyethylene films by homogeneous catalytic ozonation. *Chem. Eng. Res. Des.* **2012**, *90*, 1800–1806. [CrossRef]
45. Ameta, R.; Chohadia, A.K.; Jain, A.; Punjabi, P.B. Fenton and Photo-Fenton Processes. In *Advanced Oxidation Processes for Waste Water Treatment: Emerging Green Chemical Technology*; Ameta, S.C., Ameta, R., Eds.; Academic Press: Chennai, India, 2018; pp. 49–87. [CrossRef]
46. Wang, N.; Zheng, T.; Zhang, G.; Wang, P. A review on Fenton-like processes for organic wastewater treatment. *J. Environ. Chem. Eng.* **2016**, *4*, 762–787. [CrossRef]
47. Ribas Fargas, D. In Situ Groundwater Remediation Treatments: Natural Denitrification Study and Nano Zero Valent Iron Production. Ph.D. Thesis, Universitat Politècnica de Catalunya, Barcelona, Spain, 2017.
48. Jang, M.-H.; Lim, M.; Hwang, Y.S. Potential environmental implications of nanoscale zero-valent iron particles for environmental remediation. *Environ. Health Toxicol.* **2014**, *29*, e2014022. [CrossRef]
49. Gacitua, M.; Pavez, L.; Escudey, M.; Antilén, M. Adsorption of Zerovalent Iron Nanoparticles in the Inorganic Fraction of Volcanic Soils. *J. Soil Sci. Plant. Nutr.* **2022**, *22*, 2392–2405. [CrossRef]
50. Kušić, H.; Lončarić Božić, A.; Koprivanac, N. Fenton type processes for minimization of organic content in coloured wastewaters: Part I: Processes optimization. *Dyes Pigment.* **2007**, *74*, 380–387. [CrossRef]
51. Fentons Reagent General Chemistry Using H_2O_2. Available online: https://www.h2o2.com/pages.aspx?pid=143&name=General-Chemistry-of-Fenton-s-Reagent (accessed on 7 July 2023).
52. Eyjolfsson, R. Chapter One—Introduction. In *Design and Manufacture of Pharmaceutical Tablets*; Eyjolfsson, R., Ed.; Academic Press: London, UK, 2015; pp. 1–28. [CrossRef]
53. Miloloža, M.; Ukić, Š.; Cvetnić, M.; Bolanča, T.; Kučić Grgić, D. Optimization of Polystyrene Biodegradation by *Bacillus cereus* and *Pseudomonas alcaligenes* Using Full Factorial Design. *Polymers* **2022**, *14*, 4299. [CrossRef]
54. Bule Možar, K.; Miloloža, M.; Martinjak, V.; Cvetnić, M.; Ocelić Bulatović, V.; Mandić, V.; Bafti, A.; Ukić, Š.; Kučić Grgić, D.; Bolanča, T. Bacteria and Yeasts Isolated from the Environment in Biodegradation of PS and PVC Microplastics: Screening and Treatment Optimization. *Environments* **2023**, *10*, 207. [CrossRef]
55. Cvetnic, M.; Ukic, S.; Kusic, H.; Bolanca, T.; Loncaric Bozic, A. Photooxidative Degradation of Pesticides in Water; Response Surface Modeling Approach. *J. Adv. Oxid. Technol.* **2017**, *20*, 20160172. [CrossRef]

56. Markic, M.; Cvetnic, M.; Ukic, S.; Kusic, H.; Bolanca, T.; Loncaric Bozic, A. Influence of process parameters on the effectiveness of photooxidative treatment of pharmaceuticals. *J. Environ. Sci. Health A* **2018**, *53*, 338–351. [CrossRef] [PubMed]
57. Miloloža, M.; Bule, K.; Ukić, Š.; Cvetnić, M.; Bolanča, T.; Kušić, H.; Ocelić Bulatović, V.; Kučić Grgić, D. Ecotoxicological Determination of Microplastic Toxicity on Algae *Chlorella* sp.: Response Surface Modeling Approach. *Water Air Soil Pollut.* **2021**, *232*, 327. [CrossRef]
58. Liu, X. *Organic Chemistry I*; Kwantlen Polytechnic University: Surrey, BC, Canada, 2021; p. 198.
59. Celina, M.C.; Linde, E.; Martinez, E. Carbonyl Identification and Quantification Uncertainties for Oxidative Polymer Degradation. *Polym. Degrad. Stab.* **2021**, *188*, 109550. [CrossRef]
60. Kumari, A.; Chaudhary, D.R.; Jha, B. Destabilization of polyethylene and polyvinylchloride structure by marine bacterial strain. *Environ. Sci. Pollut. Res.* **2019**, *26*, 1507–1516. [CrossRef]
61. Subramani, M.; Sepperumal, U. FTIR analysis of bacterial mediated chemical changes in Polystyrene foam. *Ann. Biol. Res.* **2016**, *7*, 55–61.
62. Lu, J.; Ma, S.; Gao, J. Study on the Pressurized Hydrolysis Dechlorination of PVC. *Energy Fuels* **2002**, *16*, 1251–1255. [CrossRef]
63. Dobberphul, H. *Organic Chemistry—MLC*; Martin Luther College: New Ulm, MN, USA; p. 3.1.12.1. Available online: https://chem.libretexts.org/Courses/Martin_Luther_College/Organic_Chemistry_-_MLC/03:_Alcohols_Ethers_Thiols_Sulfides_and_Amines/3.01:_Alcohols_and_Phenols/3.1.12:_Spectroscopy_of_Alcohols_and_Phenols (accessed on 16 February 2024).
64. Gaumet, S.; Gardette, J.-L. Photo-oxidation of poly(vinyl chloride): Part 2—A comparative study of the carbonylated products in photo-chemical and thermal oxidations. *Polym. Degrad. Stab.* **1991**, *33*, 17–34. [CrossRef]
65. Ahmed, A.; El-Hiti, G.A.; Hadi, A.G.; Ahmed, D.S.; Baashen, M.A.; Hashim, H.; Yousif, E. Photostabilization of Poly(vinyl chloride) Films Blended with Organotin Complexes of Mefenamic Acid for Outdoor Applications. *Appl. Sci.* **2021**, *11*, 2853. [CrossRef]
66. Zhang, Y.; Pedersen, J.N.; Eser, B.E.; Guo, Z. Biodegradation of polyethylene and polystyrene: From microbial deterioration to enzyme discovery. *Biotechnol. Adv.* **2022**, *60*, 107991. [CrossRef]
67. Ainali, N.M.; Bikiaris, D.N.; Lambropoulou, D.A. Aging effects on low- and high-density polyethylene, polypropylene and polystyrene under UV irradiation: An insight into decomposition mechanism by Py-GC/MS for microplastic analysis. *J. Anal. Appl. Pyrolysis* **2021**, *158*, 105207. [CrossRef]
68. Yousif, E.; Hasan, A. Photostabilization of poly(vinyl chloride)—Still on the run. *J. Taibah Univ. Sci.* **2015**, *9*, 421–448. [CrossRef]
69. Payne, L.B. The Dehydrochlorination Mechanism of the Internal Allylic Chloride Structure in Poly(vinyl chloride). Ph.D. Thesis, College of William & Mary, Williamsburg, VA, USA, 2020. [CrossRef]
70. Li, X.; Wang, J.; Rykov, A.I.; Sharma, V.K.; Wei, H.; Jin, C.; Liu, X.; Li, M.; Yu, S.; Sun, C.; et al. Prussian blue/TiO_2 nanocomposites as a heterogeneous photo-Fenton catalyst for degradation of organic pollutants in water. *Catal. Sci. Technol.* **2015**, *5*, 504–514. [CrossRef]
71. Kowalska, B.; Klepka, T.; Kowalski, D. Influence of chlorinated water on mechanical properties of polyethylene and polyvinyl chloride pipes. *WIT Trans. Built Environ.* **2016**, *165*, 63–74. [CrossRef]
72. Jia, X.; Qin, C.; Friedberger, T.; Guan, Z.; Huang, Z. Efficient and selective degradation of polyethylenes into liquid fuels and waxes under mild conditions. *Sci. Adv.* **2016**, *2*, e1501591. [CrossRef]
73. Burelo, M.; Hernández-Varela, J.D.; Medina, D.I.; Treviño-Quintanilla, C.D. Recent developments in bio-based polyethylene: Degradation studies, waste management and recycling. *Heliyon* **2023**, *9*, e21374. [CrossRef]
74. Ul-Hamid, A.; Soufi, K.Y.; Al-Hadhrami, L.M.; Shemsi, A.M. Failure investigation of an underground low voltage XLPE insulated cable. *Anti-Corros. Method. M.* **2015**, *62*, 281–287. [CrossRef]
75. Li, T.; Zhao, P.; Lei, M.; Li, Z. Understanding Hydrothermal Dechlorination of PVC by Focusing on the Operating Conditions and Hydrochar Characteristics. *Appl. Sci.* **2017**, *7*, 256. [CrossRef]
76. Yaseen, A.A.; Yousif, E.; Al-Tikrity, E.T.B.; El-Hiti, G.A.; Kariuki, B.M.; Ahmed, D.S.; Bufaroosha, M. FTIR, Weight, and Surface Morphology of Poly(vinyl chloride) Doped with Tin Complexes Containing Aromatic and Heterocyclic Moieties. *Polymers* **2021**, *13*, 3264. [CrossRef]
77. Kumari, P.; Kumar, A. ADVANCED OXIDATION PROCESS: A remediation technique for organic and non-biodegradable pollutant. *Results Surf. Interfaces* **2023**, *11*, 100122. [CrossRef]
78. Ma, Y.-B.; Xie, Z.-Y.; Hamid, N.; Tang, Q.-P.; Deng, J.-Y.; Luo, L.; Pei, D.-S. Recent advances in micro (nano) plastics in the environment: Distribution, health risks, challenges and future prospects. *Aquat. Toxicol.* **2023**, *261*, 106597. [CrossRef] [PubMed]

water

Article

Alkaline Chemical Neutralization to Treat Acid Mine Drainage with High Concentrations of Iron and Manganese

Pingping Zhao [1,2], Ruiming Zhang [1,2,*] and Mengdi Hu [1,2]

1 Fujian Province Colleges and University Engineering Research Center of Solid Waste Resource Utilization, Longyan 364012, China; 82018039@lyun.edu.cn (P.Z.); hudadi35@163.com (M.H.)
2 College of Chemistry and Materials, Longyan University, Longyan 364012, China
* Correspondence: zhangruim07@163.com

Abstract: Due to its high acidity and toxic metal content, acid mine drainage (AMD) needs to be properly treated before being discharged into the environment. This study took the AMD collected from one specific mine in China as a sample and investigated the treatment methodology for AMD. The water quality of the AMD was measured, and the sample was treated with caustic soda (NaOH) and shell powder (one kind of conventional neutralizer, mainly composed of $CaCO_3$) by the neutralization method. The results show that the AMD has a relatively low pH (2.16) and contains high concentrations of Fe (77.54 g/L), Mn (621.29 mg/L), Cu (6.54 mg/L), Ca (12.39 mg/L), and Mg (55.04 mg/L). NaOH was an effective neutralizer to treat the AMD and performed much better than shell powder. Various metals were precipitated, in the order of Fe(III), Cu, Fe(II), Mn, Ca, and Mg. The metal removal mechanisms included precipitation, adsorption, and co-precipitation. The optimal reaction conditions were the reaction duration was selected as 5 min and the mass ratio of NaOH to AMD was 0.16:1 (*w*:*v*). By this stage, the pH rapidly increased from 2.16 to 8.53 during AMD-NaOH interactions and various metals were efficiently removed (from 86.71% to 99.99%) by NaOH. The residual mass concentrations of Fe, Mn, Cu, Ca, and Mg after the treatment were 1.52, 1.77, 0.10, 1.65, and 2.17 mg/L, respectively. These data revealed that NaOH was a good treatment regent for this kind of AMD, based on the discharge criteria of China (GB28661 2012). Also, the shell powder was a helpful neutralizer for pH adjustment and copper removal. This neutralization method has the advantages of convenient operation, high speed, good effect, simple equipment, and low infrastructure cost. In addition, the resulting neutralized residue is a valuable and high-quality raw material, which can be used in metal smelting and separation.

Keywords: acid mining drainage (AMD); heavy metals; neutralization; precipitation; NaOH; shell powder

1. Introduction

Acid mine drainage (AMD) is one kind of typical wastewater mostly produced in the mining industry [1]. It is mainly derived from metal sulfide minerals present in active or abandoned mines, tailings ponds, waste rocks, and sulfuric acid soils [2–4]. Among these sulfides, pyrite (FeS_2) is the most important because of its ease of oxidation [5–7]. In addition, pyrrhotite ($Fe_{1-x}S$, where $0 < x < 0.2$) and arsenopyrite (FeAsS) are also common metal sulfide minerals that generate AMD [8]. When exposed to oxygen (O_2), water, and certain microorganisms (such as Thiobacillus thiooxidans and Thiobacillus ferrooxidans), these sulfides undergo complex chemical and biological oxidation, resulting in the production of AMD [2,4,9].

The main environmental hazards posed by AMD are its extremely low pH (usually below 3) and high concentrations of toxic metals ions [10,11]. A low pH is hazardous to the survival of living organisms and may lead to metals leaching from rocks into aquatic systems. That is to say, as water becomes acidic, the mobility and solubility of heavy metals increase, leading to the chemical, physical, and biological erosion of the soil and water bodies [4,12]. Due to their toxicity, non-degradation, and bioaccumulation, in the long run, heavy metals in AMD can accumulate in the food chain, posing a significant threat to microorganisms, plants, animals, and humans [13–15]. For example, large amounts of iron (Fe) and manganese (Mn) in AMD entering water bodies can cause various symptoms in the local residents, such as drowsiness, muscle tremor, hepatic necrosis, shock, metabolic acidosis, DNA damage, and in severe cases, even death [16]. In summary, if AMD is directly discharged without treatment, it may pose a serious threat to the environment, especially the soil, surface water, groundwater, and aquatic communities [17–19]. Therefore, it is necessary to properly treat AMD before discharge to meet the required discharge standard.

The AMD treatment includes active and passive treatment approaches [8]. Active treatment methods commonly include neutralization, precipitation, adsorption, ion exchange, membrane separation, and electrochemical approach [20–24]. Passive treatment methods include constructed wetlands, limestone drains and channels, permeable reactive barriers, and sulfate reduction bioreactors [25–29]. Each of these treatment methods has its unique advantages and disadvantages when dealing with AMD. In terms of the AMD treatment methods, active treatments can be used to treat various types of AMD through a single or combination of multiple methods [11]. However, active treatments require continuous chemical addition and secondary pollutant treatment [12]. The advantages of passive treatments are their low cost of implementation, operation, maintenance, and management [28]. Their disadvantages are their low acid load, prolonged process time, poor effectivity during winter, unstable water output, and being less efficient over time [30]. Among these methods, chemical neutralization is currently the most common and widespread method for treating AMD due to its rapid reaction rate and high removal efficiency [13]. The commonly utilized alkaline neutralizing agents include lime and limestone ($CaCO_3$), calcium oxide (CaO), hydrated lime ($Ca(OH)_2$), soda ash (Na_2CO_3), caustic soda (NaOH), ammonium hydroxide (NH_4OH), and magnesium hydroxide ($Mg(OH)_2$) [13]. Among these neutralizing agents, calcium-based reagents (particularly limestone) are used extensively worldwide because of their low cost, abundant reserves, and convenient storage [16]. However, using limestone or hydrated lime to treat AMD usually generates a large amount of metal-rich sludge, which is hard to settle, recycle, and reuse [31]. To summary, although the treatment cost of AMD using limestone or hydrated lime is cheap, the disposal cost of the sludge is relatively expensive [32]. In recent years, alkaline industrial by-products, such as concrete fines, lime kiln dust, coal ash, steel mill slag, red mud, and fly ash, have replaced traditional alkaline chemicals for treating AMD [33–36]. Using alkaline industrial by-products further reduces the treatment cost, but also makes it harder to reuse the sludge because industrial by-products introduce a variety of cations and anions into the sludge. Therefore, more effective and less sludge-producing AMD treatment processes are required. Particularly, if the resulting solids by the neutralization method are relatively pure and free of complex impurity components, they can be conveniently used as high-quality raw materials for subsequent metal smelting.

In this study, the actual AMD derived from one mine in China was treated by the neutralization method. The water quality of the actual AMD, including the pH and various metal ions' concentrations, was measured first. Based on the above results, NaOH was selected as the neutralizing agent to treat the AMD. The performance of NaOH on raising the pH and removing metal ions was investigated and compared to the performance of shell powder (one conventional neutralization material). Then, the optimal dosage and neutralization time of NaOH were investigated. Finally, characterization was performed with X-ray diffraction (XRD) and Fourier transform infrared spectroscopy (FTIR) for the neutralized residues. The research results can provide data support for operating conditions' selection and process adjustment in the actual production.

2. Materials and Methods

2.1. The Acid Mining Drainage and Reagents

The actual acid mine drainage (AMD) in this study was collected from one mine in China just before entering the AMD treatment plant, packed in clean 20 L PVC plastic drums, transported to the laboratory, and stored at 4 °C until analysis and treatment.

All chemical reagents used in this study were analytically pure and used as received. Hydrochloric acid (purity 37%) and NaOH were purchased from Xilong Scientific Co., Ltd. (Chengdu, China), and other reagents, for example, standard metal solutions (1000 mg/L), were purchased from Macklin Biochemical Technology Co., Ltd. (Shanghai, China). Distilled water was used for all experiments.

2.2. Methodology

2.2.1. Characterization of the Acid Mine Drainage

Referring to the Chinese wastewater discharge standard of "Emission standard of pollutants for mining and mineral processing industry" (GB28661-2012) [11], the water pollution characteristics of the AMD were analyzed, including the pH value and various metal concentrations. The pH was measured using a pH meter (PHS-3C, Ramag, Shanghai, China). The concentrations of metals (Zn, Cu, Mn, Fe, Cd, Cr, Pb, Ni, Ag, Ca, and Mg) were measured by a flame atomic absorption spectrophotometer (Shimadzu AA-7000F, Kyoto, Japan).

2.2.2. Neutralization Performance Experiments of NaOH and Shell Powder

The chemical neutralization method was used to treat the above raw AMD. The experiments were conducted at room temperature (25 ± 3 °C). The neutralization performance of NaOH and shell powder were investigated by batch experiments. The neutralizer NaOH (0.5, 1.0, 1.5, 2.0, and 2.5 g) or shell powder (0.5, 1.0, 1.5, 2.0, 2.5 g, and 3.0 g) with different dosage ranges and 10 mL AMD solution were added to a series of 50 mL centrifuge tubes. Subsequently, all tubes were placed in a Rotary Mixer (QB-228, Haimen Kylin-Bell Lab Instruments Co., Ltd., Haimen, China) for 2 h at 60 rpm. After the reaction, the pH of the suspension was measured. The remaining suspension was centrifuged at 4000 rpm for 5 min and filtered through 0.45 μm PTFE syringe-tip filters. The residual metal concentrations (Fe, Mn, Cu, Ca, and Mg) were analyzed by a flame atomic absorption spectrophotometer. The removal rate of each metal element was calculated by dividing the concentration difference before and after treatment by the concentration before treatment [11].

2.2.3. The Optimal Dosage Experiments of NaOH

According to the experimental results of Section 2.2.2, the optimum dosage of NaOH is between 1.5 g and 2.0 g. Therefore, to further explore the optimal dosage of NaOH, the neutralization experimental process was conducted once again. Therefore, in order to neutralize 10 mL AMD, the dosages of NaOH were selected as 1.6, 1.7, 1.8, and 1.9 g. The neutralization reaction time was set to 2 h.

2.2.4. The Optimal Neutralization Time Experiments of NaOH

The influence of the neutralization reaction time of NaOH on the AMD treatment was determined through batch experiments. A total of 2.0 g NaOH and 10 mL AMD solution were added to a series of 50 -mL centrifuge tubes. The neutralization experiments were conducted under different neutralization reaction times (1, 2, 3, 5, 7, 10, 15, 20, 30, 45, 60, 90, 120, 180, and 240 min).

2.2.5. Characterization of the Neutralized Residues

The neutralized residues were placed in an oven (DHG-9076A, Shanghai Kunquan biotechnical Co., Ltd., Shanghai, China) at 50 °C for 48 h, and then ground into a powder. The mineral phases of the residues were performed by a powder X-ray diffraction spectroscopy (XRD, Dandong Fang-yuan Instruments Co., Ltd., Dandong, China) equipment, with Cu Kα scanning range from 5° to 80° (2θ) at a scan speed of 2°/min. The surface functional groups and valence bounds were determined using Fourier transform infrared spectroscopy (FTIR, Nicolet 360FT-IR spectrometer, Thermo Nicolet Corporation, Madison, WI, USA), over a wavenumber ranging from 4000 to 400 cm^{-1} by using KBr as a reference.

3. Results and Discussion

3.1. Water Quality of the Raw Acid Mine Drainage

The water quality of raw AMD before treatment and the discharge standard ("Emission standard of pollutants for mining and mineral processing industry" (GB28661-2012)) are shown in Table 1. As expected, the AMD had a relatively high acidity, with a pH of 2.16, and contained various metal elements. In the extensive literature about AMD, both actual and synthetic AMD generally contain large amounts of iron (Fe) and manganese (Mn) elements. The concentration range of Fe is generally from tens to hundreds of milligrams per liter, while the concentration of Mn is tens of milligrams per liter [11,18,23,28]. The AMD in this study had an iron concentration of up to 77.54 g/L and a Mn concentration of up to 621.29 mg/L, both of which are much higher than those in general AMD. In addition to its strong acidity, the main feature of this AMD was its high contents of Fe and Mn elements. Magnesium (Mg), calcium (Ca), and copper (Cu) were found in moderate concentrations, ranging from a few milligrams per liter to tens of milligrams per liter, while lead (Pb) was a trace element at a concentration of tens of micrograms per liter. Heavy metals, such as Zn, Cd, Cr, Ni, and Ag, were not detected in the raw AMD; so, these metals were considered no longer of concern and were not measured in the subsequent neutralization experiments.

Table 1. The water quality of the raw AMD and the relative discharge standard "Emission standard of pollutants for mining and mineral processing industry" (GB28661-2012).

Parameters	Acid Mine Drainage (AMD)	GB28661-2012
pH	2.16	6–9
Zn (mg/L)	undetected	2.0
Cu (mg/L)	6.54	0.5
Mn (mg/L)	621.29	2.0
Fe (mg/L)	77.54 (g/L)	5.0
Cd (mg/L)	undetected	0.1
Cr (mg/L)	undetected	1.5
Pb (mg/L)	0.06	1.0
Ni (mg/L)	undetected	1.0
Ag (mg/L)	undetected	0.5
Ca (mg/L)	12.39	/
Mg (mg/L)	55.04	/

According to the "Emission standard of pollutants for mining and mineral processing industry" (GB28661-2012) standard, the concentration of Pb (0.06 mg/L) in the raw AMD was below the limit value (1.0 mg/L); so, Pb was not measured in the subsequent neutralization experiments. The concentrations of Fe, Mn, and Cu in the AMD must be reduced to meet the standard requirements. Meanwhile, the acidity of the AMD must be neutralized to a natural state. Due to the high content of Mn in the AMD and the difficulty of Mn precipitation when the pH value is below 6 [13,37], the highly alkaline neutralizer NaOH was selected to treat the AMD. For comparison, shell powder (mainly composed of $CaCO_3$) was selected as a conventional neutralizer to evaluate the treatment efficiency of NaOH on the AMD. Despite Ca and Mg concentrations not being addressed in the emission standard GB28661-2012, the concentrations of these two elements are related to the hardness of water; so, they were tested in the following neutralization experiments.

3.2. Performances of NaOH and Shell Powder during the Acid Mine Drainage Treatment

The performances of NaOH and the shell powder during the AMD treatment are presented in Figure 1. As shown in Figure 1a,b, with the increase in the dosage of the two neutralizers, the pH value and the removal rates of the various metal elements increased. For NaOH, as the dosages increased, the pH of the solution rose from 2.16 to 13.77. For the shell powder, the pH of the solution quickly reached a maximum of 5.88 and remained stable. The removal rates varied with the metals. In general, NaOH had a better performance than the shell powder for the removal of various metals.

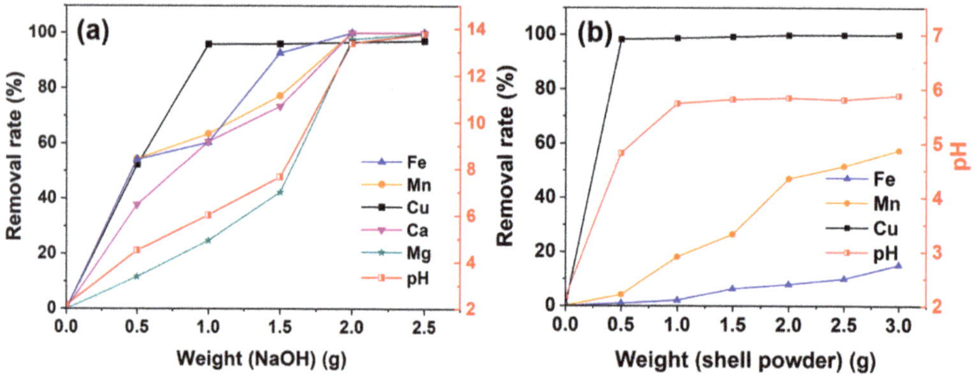

Figure 1. Performances of NaOH (**a**) and shell powder (**b**) on the AMD treatment (experimental conditions: 10 mL AMD solution; reaction time = 2 h).

In the case of Cu, when adding 1.0 g NaOH, the pH of the solution increased from 2.16 to 6.02, and the removal rate of Cu rapidly increased to over 95%. When adding 0.5 g shell powder, the pH of the solution increased from 2.16 to 4.82, and the removal rate of Cu was close to 100%. As shown in Figure 1b, the alkalinity produced by the shell powder is mainly used to precipitate Cu, while the precipitations produced by iron and Mn are very small. If only considering the removal of Cu in the AMD, the shell powder is a good neutralizing agent.

In the case of iron, when adding 0–2.5 g NaOH, the removal rate of iron showed a three-stage trend. The removal rate of Fe increased rapidly (adding 0–0.5 g NaOH, pH = 2.16–4.50) at first, then increased slowly (adding 0.5–1.0 g NaOH, pH = 4.50–6.02), and finally rapidly raised (adding 1.0–2.5 g NaOH, pH > 6.02) to near 100%. The final removal rate of iron by NaOH was higher than 99.99%. According to Hu et al., at 20 °C, the pH range of Fe^{3+} from the beginning to complete precipitation is 1.9–3.2 and the pH range of Fe^{2+} precipitation is 7.9–9.0 [38]. Therefore, it can be inferred that both Fe(III) and Fe(II) existed in the AMD. In Figure 1b, using the shell powder as the neutralizer (pH = 2.16–5.88), only less than 20% of iron was precipitated, indicating that iron in the AMD mainly existed in the form of Fe(II).

In the case of Mn, in Figure 1b, the removal rate of Mn increased with the addition of the shell powder, and its final removal rate was lower than 60%. This low removal rate was due to the low alkalinity (pH < 6.0) provided by the shell powder, resulting in only a small amount of Mn hydroxide being formed. This is consistent with the results of previous studies. It is difficult to remove Mn from the AMD by precipitation under low pH (<6.0) conditions [13,37]. In Figure 1a, similar to iron, the removal rate of Mn also showed a three-stage trend. Furthermore, the Mn removal rate in Figure 1a, when the pH is equal to 4.50 (adding 0.5 g NaOH), is much higher than that in Figure 1b, when pH is equal to 4.82 (adding 0.5 g shell powder). The difference between Figure 1a and 1b is that using the shell powder as a neutralizing agent results in a lower removal rate of iron, that is, less iron (oxy)hydroxide was formed. Therefore, it can be inferred that, in addition to the formation of manganese hydroxide precipitation, the adsorption and co-precipitation of Mn by iron (oxy)hydroxide precipitation formed by Fe(III) were also important mechanisms for Mn removal.

In the case of Mg, as the pH value increased, the removal rate of Mg continued to increase. Especially when the pH was greater than 7.65 (adding 1.5 g NaOH), the removal rate of Mg increased rapidly. These results can be interpreted by the fact that Mg has a high mobility in a solution [28] and can only form hydroxide and carbonate deposits under high pH (pH > 6.0) conditions [13,37].

In the case of Ca, when the dose of the neutralizer NaOH increased from 0 g to 2.0 g, the removal rate of Ca gradually increased and was close to 100% at the dose of 2.0 g NaOH.

How do multiple metal ions compete for OH^- to form precipitates? After adding the neutralizers, the pH value of the AMD increased, and the metal ions precipitated in the form of hydroxides, such as $Fe(OH)_2$, $Fe(OH)_3$, $Mn(OH)_2$, $Cu(OH)_2$, and $Mg(OH)_2$. Ca^{2+} could be precipitated as gypsum ($CaSO_4{\cdot}2H_2O$), bassanite ($2CaSO_4{\cdot}H_2O$), anhydrite ($CaSO_4$), $Ca(OH)_2$, and so on. Some studies stated that the main product of Ca precipitation was gypsum [20,39]. The order in which the metal ions form deposits depends on the value of the solubility product (K_{SP}). Metal ions with a small solubility product constant are more likely to deposit. Precipitation reaction equations of various metals and the K_{SP} of the precipitations at 298.15 K are shown in Table 2. According to the metals' concentrations and the discharge standard of AMD, the pH ranges of the various metals from the beginning of the precipitation to the effluent concentrations meeting the discharge standard can be calculated, as shown in Table 2. Therefore, the order in which the metals deposited was $Fe(OH)_3$, $Cu(OH)_2$, $Fe(OH)_2$, $Mn(OH)_2$, $Mn(OH)_2$, and $Mg(OH)_2$.

Table 2. Precipitation reaction equations, K_{SP}, and the pH ranges of the various metals from the beginning of the precipitation to the effluent concentrations reaching the discharge standard.

Metals	Concentrations	GB28661-2012	Precipitation Reaction Equations	K_{SP} [40]	Ranges of pH for Metal Precipitation
Fe	77.54 g/L (1.39 mol/L)	5.0 mg/L (8.95×10^{-5} mol/L)	$Fe^{3+} + 3OH^- = Fe(OH)_3\downarrow$ [18]	4.0×10^{-38}	1.5~2.9
			$Fe^{2+} + 2OH^- = Fe(OH)_2\downarrow$ [18]	8.0×10^{-16}	6.4~9.5
Cu	6.54 mg/L (1.03×10^{-4} mol/L)	0.5 mg/L (7.87×10^{-6} mol/L)	$Cu^{2+} + 2OH^- = Cu(OH)_2\downarrow$	2.2×10^{-20}	6.2~6.7
Mn	621.29 mg/L (1.13×10^{-2} mol/L)	2.0 mg/L (3.64×10^{-5} mol/L)	$Mn^{2+} + 2OH^- = Mn(OH)_2\downarrow$	1.9×10^{-13}	8.6~9.9
Ca	12.39 mg/L (3.09×10^{-5} mol/L)	/	$Ca^{2+} + SO_4^{2-} = CaSO_4\downarrow$ [5]	9.1×10^{-6}	/
			$Ca^{2+} + 2OH^- = Ca(OH)_2\downarrow$	5.5×10^{-6}	>13.6
Mg	55.04 mg/L (2.26×10^{-3} mol/L)	/	$Mg^{2+} + 2OH^- = Mg(OH)_2\downarrow$	1.8×10^{-11}	>10.0

If Ca was removed from the solution in the form of $CaSO_4$, $CaSO_4{\cdot}2H_2O$, and $2CaSO_4{\cdot}H_2O$, the Ca removal rate is not related to the pH of the solution, but only to the concentrations of Ca^{2+} and $SO_4{}^{2-}$. If Ca was removed in the form of $Ca(OH)_2$, $Ca(OH)_2$ precipitation cannot occur until the pH of the solution is higher than 13.6. However, the Ca removal rate continued to rise with the increase in the pH (pH = 2.16–13.77, adding 0–2.5 g NaOH), indicating that there were other removal mechanisms besides the formation of gypsum and $Ca(OH)_2$. The K_{SP} values of $CaSO_4$ and $Ca(OH)_2$ were 9.1×10^{-6} and 5.5×10^{-6}, respectively, which were higher than that of $Mg(OH)_2$, indicating $Mg(OH)_2$ precipitated earlier than $CaSO_4$ and $Ca(OH)_2$. However, in Figure 1a, Ca^{2+} begins to form deposits earlier than Mg^{2+} and has a higher removal rate than Mg^{2+}, both of which further indicate that there are other mechanisms for Ca removal besides precipitation. Therefore, the order in which metal ions form deposits is Fe^{3+}, Cu^{2+}, Fe^{2+}, Mn^{2+}, Ca^{2+}, and Mg^{2+}.

Table 3 shows the performances of NaOH and the shell powder dosages during the AMD treatment. Comparing Tables 1 and 3, it can be seen that, when the amount of NaOH added is 1.5 g, the metal concentrations in the treated AMD exceed the relative standard limit. However, when the NaOH addition amount is 2.0 g, the metal concentrations after the treatment meet the standard limit, but the pH value is too high (>9), indicating that too much NaOH was added at this stage. Therefore, the optimal dosage of NaOH is between 1.5 g and 2.0 g.

Table 3. Metal concentrations and final pH values in the AMD treated with different dosages of NaOH and the shell powder.

Materials	Dosages (g)	Metal Concentrations (mg/L)					pH
		Fe	Mn	Cu	Ca	Mg	
NaOH	0.5	35.50	283.33	3.13	7.73	48.62	4.50
	1.0	30.73	226.85	0.27	4.89	41.41	6.02
	1.5	5.63	141.67	0.26	3.30	31.75	7.65
	2.0	0	0.56	0.22	0	1.30	13.37
	2.5	0	0.54	0.19	0	0.10	13.77
Shell powder	0.5	76,800.1	595.21	0.11	/	/	4.82
	1.0	75,912.2	508.65	0.09	/	/	5.73
	1.5	72,508.8	456.48	0.05	/	/	5.81
	2.0	71,325.0	329.61	0.01	/	/	5.84
	2.5	69,697.2	301.16	0.01	/	/	5.80
	3.0	65,701.8	264.40	0	/	/	5.88

Adding the shell powder to the AMD neutralized the acidity and precipitated metal ions. However, the maximum pH value (5.88) achieved by the shell powder was lower than the standard limit (pH > 6.0). It has been reported that treating AMD with limestone can increase the pH to 6.0–7.5 [41]. Shell powder could not provide sufficient alkalinity. At this point, almost all Cu was removed, while large amounts of Mn and iron did not precipitate and still existed in the solution. Therefore, it is not suitable to use shell powder alone to treat AMD containing high concentrations of iron and Mn. Accordingly, it can be inferred that calcium-based reagents (such as $CaCO_3$, $Ca(OH)_2$, and CaO) and alkaline industrial by-products (such as concrete fines, lime kiln dust, coal ash, steel mill slag, red mud, and fly ash) cannot effectively treat this AMD. The shell powder can be used as a pretreatment to regulate the pH and in combination with stronger neutralizers. Subsequent experiments on the optimal dosage and optimal neutralization time did not use shell powder, because it was completely meaningless to do so.

3.3. The Optimal Dosage of NaOH of the Acid Mine Drainage Treatment

The dosage experiments of NaOH were repeated in the AMD treatment to determine the optimal dosage. The main cost of treating AMD comes from the amount of neutralizer that is used; so, it is important to choose the optimal dosage of NaOH. After being neutralized, the removal rate of various metal elements and the final pH value are shown in Figure 2, and the metal concentrations and the pH value of the treated AMD are shown in Table 4.

The pH is an important parameter affecting the degree of ionization of metal ions in a solution [11,42]. There was a positive correlation between the dosage of NaOH and the pH, attributed to the reaction of NaOH and AMD. The larger the amount of NaOH added, the larger the amount of heat released, and the higher the final pH of the system. When the dosage of NaOH increased from 1.6 g to 1.7 g, the solution pH increased rapidly from 8.53 to 12.33. Subsequently, as the dosage of NaOH continued to increase, the pH value increased slowly, to over 13.0.

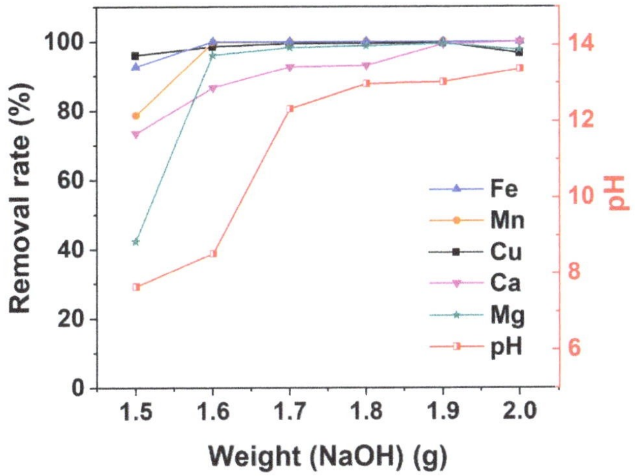

Figure 2. Effect of the NaOH dosage on the AMD treatment (experimental conditions: 10 mL AMD solution; reaction time = 2 h).

Table 4. Metal concentrations and the final pH in the treated AMD with different dosages of NaOH.

NaOH (g)	Metal Concentrations (mg/L)					Final pH
	Fe	Mn	Cu	Ca	Mg	
1.5	5.63	141.67	0.25	3.30	31.75	7.65
1.6	1.52	1.77	0.10	1.65	2.17	8.53
1.7	1.48	0.65	0.04	0.91	0.98	12.33
1.8	1.48	0.39	0.04	0.86	0.71	12.98
1.9	1.45	0.01	0.02	0.09	0.31	13.03
2.0	0	0.56	0.22	0	1.30	13.37

As shown in Figure 2, as the pH increased, the metal removal rates rose rapidly at first and then remained stable. When the addition amount of NaOH was $\geq$1.6 g, the removal rates of the various metals were relatively high, and the effluents met the relative standard (GB28661-2012) limit.

Considering the pH value and metal concentrations in the treated AMD, the cost of the additive, and the discharge standard, it was considered that adding 1.6 g NaOH to 10 mL AMD was the best ratio. That is to say, the optimized treatment mass ratio was 0.16:1 (w:v) for NaOH powder to AMD. At this stage, the removal rates of Fe, Mn, Cu, Ca, and Mg were 100.00%, 99.71%, 98.46%, 86.71%, and 96.06%, respectively. The concentrations of Fe, Mn, Cu, Ca, and Mg in the treated AMD were 1.52, 1.77, 0.10, 1.65, and 2.17 mg/L, respectively, and the pH value was 8.53.

3.4. The Optimal Neutralization Time of the Acid Mine Drainage Treatment

The effect of the neutralization time of NaOH on the AMD treatment is shown in Figure 3. All metals were completely removed (almost 100%) from the solution within 1 min, and then, all metal removal rates, except for Cu, remained stable over a treatment time of 240 min. The Cu removal rate decreased after 7 min, dropped to 97.30% at 60 min, and then remain stable. The pH value rose to 13.66 at 1 min, rose to 13.88 at 20 min, and then fluctuated slightly over a treatment time of 240 min. These results indicate that a relatively fast chemical reaction occurred. These results are consistent with previous neutralization experiments conducted with actual or synthetic AMDs. It has been reported that it takes 20 min to remove iron from AMD with $CaCO_3$, indicating that the reaction rate of $CaCO_3$ is much slower than that of NaOH [35]. The reaction time is of great practical significance for the treatment of AMD. Compared to $CaCO_3$, using NaOH as the neutralizer allows the treatment of a greater amount of AMD within the same time interval; so, a smaller neutralizing tank is required. In the actual treatment of AMD, in order to ensure a complete reaction, the optimal reaction time is recommended to be longer than 1 min, such as 5 min.

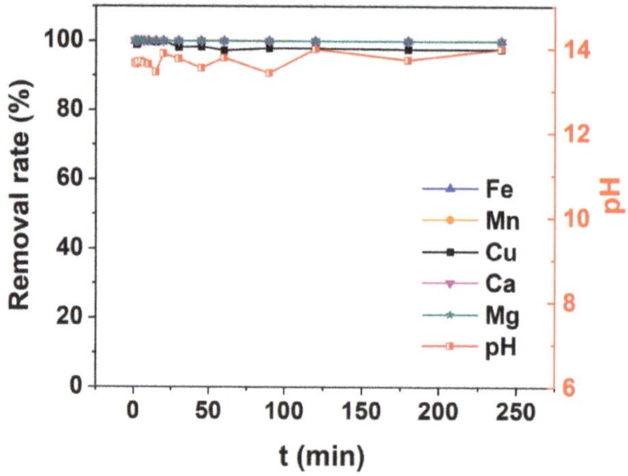

Figure 3. Effect of the neutralization time on the AMD treatment (experimental conditions: dosage of NaOH = 2.0 g; 10 mL AMD solution).

3.5. *Characterization of the Neutralized Residues*

3.5.1. XRD Analysis

The XRD patterns of the neutralized residues are presented in Figure 4. The XRD characterization of the neutralized residues showed the formation of new minerals, such as $Mn(OH)_2$ (JCPDS #73-1133), $Mg(OH)_2$ (JCPDS#07-0299), $Cu(OH)_2$ (JCPDS #13-0420) [43], and Fe_2O_3 (JCPDS #25-1402). However, no diffraction peaks of iron hydroxides are observed in Figure 4, implying that the iron hydroxides formed by neutralization precipitation are in an amorphous phase. This result is consistent with those of previous studies that showed that it was easy to form amorphous Fe hydroxide minerals, rather than those with a crystalline form, during the neutralization of AMD [41]. Based on previous studies, it was inferred that Fe was precipitated as goethite (α-FeOOH) and lepidocrocite (γ-FeOOH) as the pH increased [23]. Amorphous Fe minerals have a large surface area and high reactivity; so, they promote the adsorption and co-precipitation of other co-existing metals [44,45].

3.5.2. FTIR Analysis

The FTIR spectra of the neutralized residues are illustrated in Figure 5. The broad bands at 3471 cm^{-1} detected in all residues correspond to the O-H stretching vibration of the water molecule or hydroxides adsorbed on the surface of the precipitates [30]. The peak at 1635 cm^{-1} was assigned to the Mg-OH bending vibration, which gradually disappeared with the increase in the amount of NaOH. The bands at 942 cm^{-1} and 661 cm^{-1} are due to the Cu-OH bending vibration [46]. The band at 834 cm^{-1} is due to $CaCO_3$. The strong band at 734 cm^{-1} is due to the strong Mn-O stretching mode of $Mn(OH)_2$ [47]. Two distinct bands at 594 cm^{-1} and 538 cm^{-1} are observed, which can be attributed to Fe-O vibrational stretching [48]. The presence of Fe-based minerals confirms that the Fe species contained in the AMD were precipitated into neutralized residues.

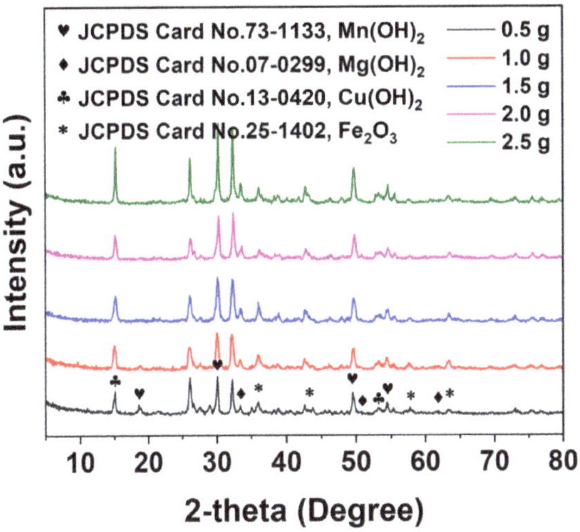

Figure 4. XRD patterns of the neutralization residues.

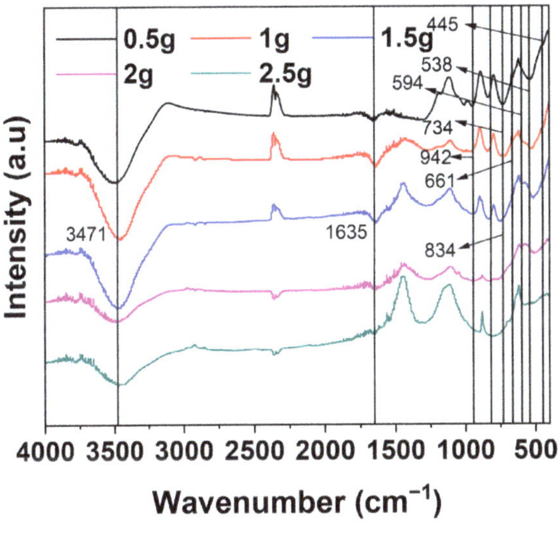

Figure 5. FTIR spectra for several neutralized residues.

4. Conclusions

The raw AMD had a relatively low pH (2.16) and contained various metals, such as Fe (77.54 g/L), Mn (621.29 mg/L), Cu (6.54 mg/L), Ca (12.39 mg/L), and Mg (55.04 mg/L). Based on the water quality of the raw AMD, neutralizing methods, including the neutralizers NaOH and shell powder, were employed in batch tests to treat the AMD. The results show that NaOH performed much better than the shell powder in increasing the pH and removing metals. Various metals were removed by NaOH, in the order of Fe(III), Cu, Fe(II), Mn, Ca, and Mg. The shell powder was a helpful neutralizer used for pH adjustment and Cu removal. However, it could not completely treat this AMD containing high concentrations of iron and Mn by itself. Precipitation, adsorption, and co-precipitation are important mechanisms for metal removal. In particular, iron (oxy)hydroxide could greatly promote the removal of Mn, Ca, and Mg. The optimum treatment mass ratio of NaOH to AMD was 0.16:1 (*w*:*v*) and the optimum reaction time was 5 min. At this point, the pH in the treated AMD was 8.53, and the Fe, Mn, Cu, Ca, and Mg concentrations after treatment were 1.52, 1.77, 0.10, 1.65, and 2.17 mg/L, respectively. The water quality after the neutralization reached the discharge standard ("Emission standard of pollutants for mining and mineral processing industry" (GB28661-2012)).

Author Contributions: Conceptualization, R.Z.; methodology, R.Z. and P.Z.; formal analysis, P.Z.; investigation, M.H.; resources, R.Z. and M.H.; data curation, P.Z. and M.H.; writing—original draft preparation, P.Z.; writing—review and editing, R.Z.; visualization, P.Z.; project administration, R.Z.; funding acquisition, P.Z. All authors have read and agreed to the published version of the manuscript.

Funding: This research was funded by the Natural Science Foundation of Fujian Province, China (No. 2022J05250) and the Qimai Natural Science Foundation of Longyan City, China (No. XLQM002).

Data Availability Statement: The original contributions presented in the study are included in the article materials; further inquiries can be directed to the corresponding author.

Acknowledgments: We would like to thank the editors and reviewers for their suggestions that improved the scientific soundness of this manuscript.

Conflicts of Interest: The authors declare no conflicts of interest.

References

1. Daraz, U.; Li, Y.; Ahmad, I.; Iqbal, R.; Ditta, A. Remediation technologies for acid mine drainage: Recent trends and future perspectives. *Chemosphere* **2023**, *311*, 137089. [CrossRef] [PubMed]
2. Naidu, G.; Ryu, S.; Thiruvenkatachari, R.; Choi, Y.; Jeong, S.; Vigneswaran, S. A critical review on remediation, reuse, and resource recovery from acid mine drainage. *Environ. Pollut.* **2019**, *247*, 1110–1124. [CrossRef] [PubMed]
3. Tabelin, C.B.; Veerawattananun, S.; Ito, M.; Hiroyoshi, N.; Igarashi, T. Pyrite oxidation in the presence of hematite and alumina: I. Batch leaching experiments and kinetic modeling calculations. *Sci. Total Environ.* **2017**, *580*, 687–698. [CrossRef] [PubMed]
4. Ighalo, J.O.; Kurniawan, S.B.; Iwuozor, K.O.; Aniagor, C.O.; Ajala, O.J.; Oba, S.N.; Iwuchukwu, F.U.; Ahmadi, S.; Igwegbe, C.A. A review of treatment technologies for the mitigation of the toxic environmental effects of acid mine drainage (AMD). *Process Saf. Environ. Prot.* **2022**, *157*, 37–58. [CrossRef]
5. Tong, L.; Fan, R.; Yang, S.; Li, C. Development and status of the treatment technology for acid mine drainage. *Min. Metall. Explor.* **2021**, *38*, 315–327. [CrossRef]
6. Xin, R.; Banda, J.F.; Hao, C.; Dong, H.; Pei, L.; Guo, D.; Wei, P.; Du, Z.; Zhang, Y.; Dong, H. Contrasting seasonal variations of geochemistry and microbial community in two adjacent acid mine drainage lakes in Anhui Province, China. *Environ. Pollut.* **2021**, *268*, 115826. [CrossRef]
7. Wang, Z.; Xu, Y.; Zhang, Z.; Zhang, Y. Review: Acid mine drainage (AMD) in abandoned coal mines of Shanxi, China. *Water* **2021**, *13*, 8. [CrossRef]
8. Park, I.; Tabelin, C.B.; Jeon, S.; Li, X.; Seno, K.; Ito, M.; Hiroyoshi, N. A review of recent strategies for acid mine drainage prevention and mine tailings recycling. *Chemosphere* **2019**, *219*, 588–606. [CrossRef]
9. Sulonen, M.L.K.; Baeza, J.A.; Gabriel, D.; Guisasola, A. Optimisation of the operational parameters for a comprehensive bioelectrochemical treatment of acid mine drainage. *J. Hazard. Mater.* **2021**, *409*, 124944. [CrossRef]
10. Alvarenga, P.; Guerreiro, N.; Simões, I.; Imaginário, M.J.; Palma, P. Assessment of the environmental impact of acid mine drainage on surface water, stream sediments, and macrophytes using a battery of chemical and ecotoxicological indicators. *Water* **2021**, *13*, 1436. [CrossRef]
11. Huang, J.; Dong, Z.; Fu, J.; Zhang, G.; Li, C.; Tian, S.; Ning, P. Insight into the kinetic analysis of acid mine drainage treated by carbonate rock. *Environ. Technol.* **2022**, *45*, 393–403. [CrossRef]
12. Chen, G.; Ye, Y.; Yao, N.; Hu, N.; Zhang, J.; Huang, Y. A critical review of prevention, treatment, reuse, and resource recovery from acid mine drainage. *J. Clean Prod.* **2021**, *329*, 129666. [CrossRef]
13. Kefeni, K.K.; Msagati, T.A.M.; Mamba, B.B. Acid mine drainage: Prevention, treatment options, and resource recovery: A review. *J. Clean Prod.* **2017**, *151*, 475–493. [CrossRef]
14. Han, R.; Zhou, B.; Huang, Y.; Lu, X.; Li, S.; Li, N. Bibliometric overview of research trends on heavy metal health risks and impacts in 1989–2018. *J. Clean Prod.* **2020**, *276*, 123249. [CrossRef]
15. Jayarathne, A.; Egodawatta, P.; Ayoko, G.A.; Goonetilleke, A. Assessment of ecological and human health risks of metals in urban road dust based on geochemical fractionation and potential bioavailability. *Sci. Total Environ.* **2018**, *635*, 1609–1619. [CrossRef] [PubMed]
16. Jabbar Khan, A.; Akhter, G.; Ge, Y.; Shahid, M.; Rahman, K.U. Development of artificial geochemical filter to treat acid mine drainage for safe disposal of mine water in Salt Range Portion of Indus Basin—A lab to pilot scale study. *Sustainability* **2022**, *14*, 7693. [CrossRef]

17. Galhardi, J.A.; Bonotto, D.M. Hydrogeochemical features of surface water and groundwater contaminated with acid mine drainage (AMD) in coal mining areas: A case study in southern Brazil. *Environ. Sci. Pollut. Res.* **2016**, *23*, 18911–18927. [CrossRef] [PubMed]
18. Zhang, X.; Guo, J.; Hu, Q.; Gao, X.; Li, C.; Luo, M.; Wang, Y. Effects of Fe-rich acid mine drainage on percolation features and pore structure in carbonate rocks. *J. Hydrol.* **2020**, *591*, 125571. [CrossRef]
19. Kaur, G.; Couperthwaite, S.J.; Hatton-Jones, B.W.; Millar, G.J. Alternative neutralisation materials for acid mine drainage treatment. *J. Water Process Eng.* **2018**, *22*, 46–58. [CrossRef]
20. Kefeni, K.K.; Mamba, B.B. Evaluation of charcoal ash nanoparticles pollutant removal capacity from acid mine drainage rich in iron and sulfate. *J. Clean Prod.* **2020**, *251*, 119720. [CrossRef]
21. Abdullah, N.; Yusof, N.; Lau, W.J.; Jaafar, J.; Ismail, A.F. Recent trends of heavy metal removal from water/wastewater by membrane technologies. *J. Ind. Eng. Chem.* **2019**, *76*, 17–38. [CrossRef]
22. López, J.; Reig, M.; Gibert, O.; Cortina, J.L. Integration of nanofiltration membranes in recovery options of rare earth elements from acidic mine waters. *J. Clean Prod.* **2019**, *210*, 1249–1260. [CrossRef]
23. Iizuka, A.; Ho, H.-J.; Sasaki, T.; Yoshida, H.; Hayakawa, Y.; Yamasaki, A. Comparative study of acid mine drainage neutralization by calcium hydroxide and concrete sludge–derived material. *Miner. Eng.* **2022**, *188*, 107819. [CrossRef]
24. Yan, S.; Cheng, K.Y.; Morris, C.; Douglas, G.; Ginige, M.P.; Zheng, G.; Zhou, L.; Kaksonen, A.H. Sequential hydrotalcite precipitation and biological sulfate reduction for acid mine drainage treatment. *Chemosphere* **2020**, *252*, 126570. [CrossRef] [PubMed]
25. Kiiskila, J.D.; Sarkar, D.; Panja, S.; Sahi, S.V.; Datta, R. Remediation of acid mine drainage-impacted water by vetiver grass (*Chrysopogon zizanioides*): A multiscale long-term study. *Ecol. Eng.* **2019**, *129*, 97–108. [CrossRef]
26. Pat-Espadas, A.M.; Loredo Portales, R.; Amabilis-Sosa, L.E.; Gómez, G.; Vidal, G. Review of Constructed Wetlands for Acid Mine Drainage Treatment. *Water* **2018**, *10*, 1685. [CrossRef]
27. Kumar, M.; Pakshirajan, K. Continuous removal and recovery of metals from wastewater using inverse fluidized bed sulfidogenic bioreactor. *J. Clean Prod.* **2021**, *284*, 124769. [CrossRef]
28. Merchichi, A.; Hamou, M.O.; Edahbi, M.; Bobocioiu, E.; Neculita, C.M.; Benzaazoua, M. Passive treatment of acid mine drainage from the Sidi-Kamber mine wastes (Mediterranean coastline, Algeria) using neighbouring phosphate material from the Djebel Onk mine. *Sci. Total Environ.* **2022**, *807*, 151002. [CrossRef]
29. Lin, H.; Tang, Y.; Dong, Y. Construction and carbon source optimization of a microbial-plant coupled reactor for treating acid mine drainage. *Environ. Sci. Pollut. Res.* **2022**, *29*, 78862–78873. [CrossRef]
30. Xu, W.; Yang, H.; Mao, Q.; Luo, L.; Deng, Y. Removal of heavy metals from acid mine drainage by red mud–based geopolymer pervious concrete: Batch and long–term column studies. *Polymers* **2022**, *14*, 5355. [CrossRef]
31. Tolonen, E.-T.; Sarpola, A.; Hu, T.; Rämö, J.; Lassi, U. Acid mine drainage treatment using by-products from quicklime manufacturing as neutralization chemicals. *Chemosphere* **2014**, *117*, 419–424. [CrossRef] [PubMed]
32. Jiao, Y.; Zhang, C.; Su, P.; Tang, Y.; Huang, Z.; Ma, T. A review of acid mine drainage: Formation mechanism, treatment technology, typical engineering cases and resource utilization. *Process Saf. Environ. Prot.* **2023**, *170*, 1240–1260. [CrossRef]
33. Yang, M.; Cunfang, L.; Quan, X.; Cao, D. Mechanism of acid mine drainage remediation with steel slag: A review. *ACS Omega* **2021**, *6*, 30205–30213. [CrossRef] [PubMed]
34. Weinberg, R.; Coyte, R.; Wang, Z.; Das, D.; Vengosh, A. Water quality implications of the neutralization of acid mine drainage with coal fly ash from India and the United States. *Fuel* **2022**, *330*, 125675. [CrossRef]
35. Ho, H.-J.; Iizuka, A.; Vadapalli, V.R.K.; Coetzee, H.; Petrik, L.; Petersen, J.; Ojumu, T. Potential investigation of concrete fines as an alternative material: A novel neutralizer for acid mine drainage treatment. *Environ. Technol. Innov.* **2023**, *29*, 102985. [CrossRef]
36. Moyo, A.; Parbhakar-Fox, A.; Meffre, S.; Cooke, D.R. Alkaline industrial wastes—Characteristics, environmental risks, and potential for mine waste management. *Environ. Pollut.* **2023**, *323*, 121292. [CrossRef] [PubMed]
37. Choi, J.; Kwon, D.; Yang, J.S.; Lee, J.Y.; Park, Y.T. Comparison of Fe and Mn removal using treatment agents for acid mine drainage. *Environ. Technol.* **2009**, *30*, 445–454. [CrossRef] [PubMed]
38. Hu, X.; Yang, H.; Tan, K.; Hou, S.; Cai, J.; Yuan, X.; Lan, Q.; Cao, J.; Yan, S. Treatment and recovery of iron from acid mine drainage: A pilot-scale study. *J. Environ. Chem. Eng.* **2022**, *10*, 106974. [CrossRef]
39. Cheong, Y.-W.; Cho, D.-W.; Yim, G.-J.; Park, H.-S.; Kim, S.-J.; Lee, J.-H. Geochemical assessment of gypsum scale formation in the hydrated lime neutralization facility of the Daedeok Mine, South Korea. *Minerals* **2022**, *12*, 574. [CrossRef]
40. Zhong, G. *Inorganic and Analytical Chemistry*; Science Press: Beijing, China, 2014; pp. 416–417.
41. Santomartino, S.; Webb, J.A. Estimating the longevity of limestone drains in treating acid mine drainage containing high concentrations of iron. *Appl. Geochem.* **2007**, *22*, 2344–2361. [CrossRef]
42. Iakovleva, E.; Mäkilä, E.; Salonen, J.; Sitarz, M.; Wang, S.; Sillanpää, M. Acid mine drainage (AMD) treatment: Neutralization and toxic elements removal with unmodified and modified limestone. *Ecol. Eng.* **2015**, *81*, 30–40. [CrossRef]
43. Peng, C.-Y.; Hou, C.-C.; Chen, Q.-Q.; Wang, C.-J.; Lv, X.-J.; Zhong, J.; Fu, W.-F.; Che, C.-M.; Chen, Y. $Cu(OH)_2$ supported on $Fe(OH)_3$ as a synergistic and highly efficient system for the dehydrogenation of ammonia-borane. *Sci. Bull.* **2018**, *63*, 1583–1590. [CrossRef] [PubMed]

44. Ye, Z.; Zhou, J.; Liao, P.; Finfrock, Y.Z.; Liu, Y.; Shu, C.; Liu, P. Metal (Fe, Cu, and As) transformation and association within secondary minerals in neutralized acid mine drainage characterized using X-ray absorption spectroscopy. *Appl. Geochem.* **2022**, *139*, 105242. [CrossRef]
45. Johnston, C.; Chrysochoou, M. Mechanisms of chromate, selenate, and sulfate adsorption on Al-substituted ferrihydrite: Implications for ferrihydrite surface structure and reactivity. *Environ. Sci. Technol.* **2016**, *50*, 3589–3596. [CrossRef] [PubMed]
46. Toupance, T.; Kermarec, M.; Lambert, J.-F.; Louis, C. Conditions of formation of copper phyllosilicates in silica-supported copper catalysts prepared by selective adsorption. *J. Phys. Chem. B* **2002**, *106*, 2277–2286. [CrossRef]
47. Wang, X.; Andrews, L. Infrared spectra of $M(OH)_{1,2,3}$ (M = Mn, Fe, Co, Ni) molecules in solid argon and the character of first row transition metal hydroxide bonding. *J. Phys. Chem. A* **2006**, *110*, 10035–10045. [CrossRef]
48. Masindi, V.; Foteinis, S.; Chatzisymeon, E. Co-treatment of acid mine drainage and municipal wastewater effluents: Emphasis on the fate and partitioning of chemical contaminants. *J. Hazard. Mater.* **2022**, *421*, 126677. [CrossRef]

water

MDPI

Article

Microbiological Mechanisms for Nitrogen Removal Using Anaerobic Fermentation Liquid from Spent Mushroom Substrates as a Carbon Source

Ruihuan Chen [1,2,*], Weihong Zhang [3], Xiaohui Bi [3], Yan Jin [1,2] and Yunlong Yang [1,2,*]

1 College of Life and Environmental Science, Wenzhou University, Wenzhou 325035, China
2 National and Local Joint Engineering Research Center of Ecological Treatment Technology for Urban Water Pollution, Wenzhou University, Wenzhou 325035, China
3 College of Life Science, Fujian Agriculture and Forestry University, Fuzhou 350002, China
* Correspondence: rhchen@wzu.edu.cn (R.C.); longyunyang@126.com (Y.Y.)

Abstract: In wastewater treatment, a low C/N ratio highly inhibits the bioremoval of nitrogen, and commercial external carbon sources are widely used. In order to obtain an economical substitute, fermentation broth of spent mushroom substrates (SMS) was employed here as a carbon source for denitrification in a sequencing batch reactor (SBR). During the domestication process, the SMS fermentation broth-feeding treatment presented comparable nitrogen removal ability (74.44%) with a commercial carbon source group (77.99%). *Rhodobacter*, *Lactobacillus* and *Pseudomonas* were the dominant bacteria in the fermentation broth, and *Saccharomycetales Gymnopilus dilepis* was the dominant fungi. At the early domestication stage, the relatively high concentration of fermentation broth led to a much lower abundance of typical nitrate reductase genes than the control group. Furthermore, extracellular polymeric substance (EPS) formation was observed in the broth-feeding sample. The microbial structure dynamic was investigated, which showed a high influent effect when 20% fermentation broth was added. As domestication proceeded, similar dominant species in the control and broth-feeding treatments were observed. Overall, SMS fermentation broth can be used as a promising substitute to replace a costly commercial carbon source.

Keywords: nitrogen removal; agricultural waste; additional carbon source; SBR reactor

1. Introduction

Effective removal of nitrogen (N) is one of the most important targets in domestic wastewater treatment plants, and biological denitrification plays a significant role in this process. The denitrifying microbes can be divided into heterotrophic and autotrophic, with heterotrophic denitrifying bacteria being the most common denitrifying bacteria in nature as well as in activated sludge [1]. Thus, heterotrophic denitrification has been widely applied due to its low treatment cost and high efficiency [2]. However, it has already been reported that sufficient carbon sources must be supplied to ensure complete denitrification [3], while the current domestic sewage generally has a low carbon–nitrogen ratio (C/N). Consequently, an additional carbon source is necessary to increase the denitrification rate and completeness when the C/N is too low [4–7]. As yet, organic compounds such as acetate, alcohol, and glucose have been used as commercial external carbon sources, but their relatively high cost is unfavorable. Thus, it is significant to seek alternatives with lower costs. Agricultural wastes are considered an important resource and could be used as external carbon sources with high economic benefits [8,9].

Edible mushrooms are an important food resource, and their cultivation process could generate large amounts of solid waste (spent mushroom substrates (SMS)), leading to a formidable challenge for disposal management [10–12]. Besides the various recycling methods for SMS [13–19], we have previously proven that the acid hydrolysates of SMS can act as an external carbon source for the treatment of low C/N ratio wastewater using a sequencing batch reactor (SBR) [20]. However, the acid pretreatment of SMS proceeds at a high temperature with additional sulfur acid, and a more economical and convenient method should be proposed. Fermentation is a favorable option for extracting carbon sources from various biodegradable biomasses; the fermentation liquid of sludge [21] food waste [22] has already been reported as an excellent external carbon source. However, the effect of SMS fermentation liquid as an external carbon source on domestic wastewater denitrification has not been investigated. Moreover, as a biological production, using fermentation liquid would consequently introduce various different microbes into the activated sludge and probably lead to a change in the micro-community and the denitrifying function. However, there are few reports about the microbial community dynamic of activated sludge influenced by the fermentation liquid carbon source, especially the important but unstable acclimatization stage.

Thus, in this research, we used the anaerobic fermentation liquid from a spent mushroom substrate, comprising *Hypsizygus marmoreus* as the carbon source, for denitrification by the SBR process. The main objectives are: (1) to investigate the effect of SMS anaerobic fermentation liquid on the SBR nitrogen removal performance and physicochemical property of activated sludge during the acclimatization process; (2) to clarify the microbiological mechanisms of activated sludge acclimatization. Our research can provide valuable information on using SMS fermentation as an external carbon source to enhance the removal of nitrogen in wastewater.

2. Material and Methods

2.1. Preparation of SMS Fermentation Liquid

The SMS of *Hypsizygus marmoreus* was used as the raw material and was obtained from Gutian, Fujian, PR China. The SMS was dried in air and then ground to a particle size of 800 μm and was stored at 4 °C in airtight containers for further use. Fermentation liquid was prepared as follows. Briefly, 250 g ground SMS sample and 1 L tap water were mixed and added into an anaerobic fermentation cylinder (2 L) for natural fermentation at room temperature. The pressure of the cylinder was adjusted by gas emission in the first 3 days. After 14 days, the fermentation mixture was filtered with a 4-layer gauze, and then the filtrate was centrifuged at 12,000 rpm. The supernatant was stored at 4 °C for further investigation. The specific components of the fermentation liquid were subject to analysis by the Shanghai WEIPU Testing Technology Group Co., Ltd. (Shanghai, China).

2.2. SBR Device and Operation

An SBR device with 6 L working volume was employed, where the temperature was held at 30 °C and 5 effective aerobic denitrifying bacteria (*Enterobacter asburiae*, *Pseudomonas putida*, *Bacillus* sp. K5, *Acinetobacter* sp.TX5, and *Stenotrophomonas* sp.) kept in our laboratory were added into the device. Each bacteria was harvested at the late log phase and adjusted to OD_{600} = 1.0, then they were mixed with an equal volume ratio, and the total mixed bacteria solution added into each device was 300 mL (5% volume ratio). Synthetic wastewater was prepared for the treatment. In the control group, the synthetic wastewater contained $FeSO_4{\cdot}7H_2O$ (0.1 g/L), $MgSO_4{\cdot}7H_2O$ (0.2 g/L), NH_4Cl (0.4 g/L), KH_2PO_4 (0.5 g/L), $Na_2HPO_4{\cdot}12H_2O$ (1 g/L), sodium citrate (1.0 g/L), glucose (1.0 g/L), and trace element solution (2 mL/L), where glucose and ammonium chloride were used as the carbon source and nitrogen source, respectively. The per liter trace element solution contained EDTA 50.0 g, $ZnSO_4$ 2.2 g, $CaCl_2$ 5.5 g, $MnCl_2{\cdot}4H_2O$ 5.06 g, $FeSO_4$ $7H_2O$ 1.57 g, and $CoCl_2$ 1.61 g.

Regarding the test treatment, the influent was composed of 95% synthetic wastewater and 5% SMS fermentation liquid initially (calculated as the COD ratio), which was gradually increased during the operation. The SRB was started up with a 40 rpm rotation speed, and each react cycle time was about 8 h; the detailed operation parameters are shown in Table 1. The sludge was sampled at 10-day intervals, and the sampling time was fixed at the first operation cycle. The separated sludge was mixed with an equal volume of 60% glycerin and stored at −80 °C for further 16S rRNA and ITS sequencing by Sangon Biotech (Shanghai) Co., Ltd. Ammonium nitrogen (NH^{4+}-N), nitrite nitrogen (NO^{2-}-N), nitrate nitrogen (NO^{3-}-N), and COD, were measured by the standard methods [23].

Table 1. Operating parameters of the reactor.

Operation Time (d)	C/N Ratio	Aeration Rate (m^3/h)	Aeration Time (h)	Settling Time (min)	Fermentation Liquid Amount
1–5	13	20	6.5	30	5%
6–10	6.5	20	7	20	10%
11–15	6.5	20	7	10	15%
16–20	6.5	20	7	10	20%
21–25	6.5	20	7	10	25%
26–30	6.5	20	7	10	30%

2.3. High through Sequencing

The total microbial DNA of the fermentation liquid was extracted with a soil microbial DNA extraction kit according to the manufacturer's instructions (OMEGA). The V3-V4 16s rRNA and ITS1-ITS2 ITS regions were amplified with general primers, as shown in Table 2, and the sequencing was carried out by the Biomarker Biotechnology Company (Beijing, China). The data was analyzed as in our previous report [20].

Table 2. Primers for high through sequencing.

Sequencing Type	Forward Primer		Reverse Primer	
16S rDNA V3-V4	341F	ACTCCTACGGGAGGCAGCAG	805R	GACTACHVGGGTATCTAATCC
ITS ITS1-ITS2	ITS1	CTTGGTCATTTAGAGGAAGTAA	ITS2	GCTGCGTTCTTCATCGATGC

2.4. Functional Gene Quantification

An absolute quantitative polymerase chain reaction (qPCR) was employed to quantify the typical nitrite reductase genes *NirS* and *NirK* in activated sludge samples. The qPCR system and standard curve were prepared as in our previous report [20].

2.5. Extracellular Polymeric Substance Extraction and Determination

In order to investigate the effect of fermentation liquid on extracellular polymeric substances (EPS), briefly, 50 mL of sludge was collected and settled for 30 min. Then, the supernatant was discarded, and 50 mL of deionized water was added to resuspend the sludge. The sludge was separated by centrifugation at 8000 rpm for 15 min and then resuspended with 10 mL deionized water, which was heated at 65 °C for 15 min. Then, the sample was centrifuged at 12,000 rpm for 30 min, and the supernatant was stored at 4 °C for further determination. Extracellular protein (PN) and polysaccharides (PS) were determined by the Coomassie Brilliant Blue method and anthrone–sulfuric acid colorimetry [24].

2.6. Analytical Techniques and Statistical Analysis

An ultraviolet-visible spectrophotometer (UV-1801) was used to determine the ammonium nitrogen (NH^{4+}-N), nitrite nitrogen (NO^{2-}-N), nitrate nitrogen (NO^{3-}-N), and COD concentrations. The water content in the fermentation liquid was analyzed with a Karl Fischer volumetric titrator (KSQL-310S), and the total protein and total polysaccharide was determined with an ultraviolet–visible spectrophotometer. High-performance liquid chromatography (HPLC, Thermo Fisher U3000, Waltham, MA, USA) equipped with a UV detector was employed to analyze the small molecular acids, where 20 mM NaH_2PO_4 solution (80%) and acetonitrile (20%) were used as the mobile phase. For the quantification of free amino acid concentrations, liquid chromatography coupled with a triple four-pole tandem mass spectrometry (LC-MS, AB5000, SCIEX, San Jose, CA, USA) was employed. The mobile phase consisted of 0.1% formic acid solution and acetonitrile, and the temperature of the ion source was set at 450 °C with a multi-reaction monitoring mode. In order to quantify the mineral elements in the fermentation broth, inductively coupled plasma emission spectrometry (ICP-OES, ICAP 7400 Radial, Thermo Fisher, USA) was used. The fermentation broth samples were diluted (5-, 10-, 20-fold) with 5% HNO_3 solution before analysis. The significant differences were evaluated with SPSS software (Version 18.0, IBM, Armonk, NY, USA), and a 0.05 threshold value was set for the p-value. The standard error of the triplicate samples was calculated using Excel software (Version 2019, Microsoft, Redmond, WA, USA).

3. Results and Discussion

3.1. Composition and Microbial Community of Fermentation Broth

The components of fermentation liquid are shown in Table 3. The concentration of COD in the fermentation liquid was 20,000 mg/L, and the pH was 6.19, which is proper for microbes. As well as water, the main components of total protein and total polysaccharide showed similar contents, ranging from 0.4% to 0.7% and 0.5% to 0.8%, respectively. The small molecular acids in the fermentation liquid were analyzed, and acetic acid (0.33 mg/kg), glycolic acid (0.07 mg/kg), and lactic acid (0.16 mg/kg) were detected. There were 11 kinds of free amino acids detected (883.1 mg/kg total), which ranged from 12.3 to 205.1 mg/kg. It is well known that these compounds can be utilized by microbes. The fermentation liquid also contained a high concentration of mineral elements, which reached 3027.5 mg/kg, and K, Ca, and Mg accounted for about 95% of the total amount. Fe and Mg are important for the formation and stabilization of most enzymes and has also been reported for nitrite reductase [25]. Cu is important for the construction of both Cu-containing nitrite reductase (Cu-NIR) and nitrous oxide reductase [26,27]. Thus, it is suggested that influent carbon sources of fermentation broth meet the requirement well for denitrification by microbes.

Table 3. Detail of the fermentation liquid composition.

Index	Method/Instrument	Value
COD	Potassium dichromate method	20,000 mg/L
pH	pH meter	6.19
Water	Karl Fischer	96.5–98.5%
Total protein	UV-vis	0.4–0.7%
Total sugar	UV-vis	0.5–0.8%
Acetic acid	HPLC	0.33 mg/kg
Lactate	HPLC	0.07 mg/kg
Glycollic acid	HPLC	0.16 mg/kg
L-alanine	LC-MS	205.1 mg/kg
L-leucine	LC-MS	186.2 mg/kg
L-valine	LC-MS	117.8 mg/kg
L-isoleucine	LC-MS	116.3 mg/kg
L-threonine	LC-MS	79.4 mg/kg
L-proline	LC-MS	31.8 mg/kg
L-phenylalanine	LC-MS	73.5 mg/kg
L-methionine	LC-MS	25.5 mg/kg
L-aspartate	LC-MS	21.5 mg/kg
L-glycine	LC-MS	13.7 mg/kg
L-tryptophan	LC-MS	12.3 mg/kg
K	ICP-OES	1356.0 mg/kg
Ca	ICP-OES	974.0 mg/kg
Mg	ICP-OES	541.7 mg/kg

Table 3. *Cont.*

Index	Method/Instrument	Value
Na	ICP-OES	66.7 mg/kg
Si	ICP-OES	43.8 mg/kg
Cu	ICP-OES	21.9 mg/kg
Fe	ICP-OES	16.2 mg/kg
Al	ICP-OES	7.2 mg/kg

Figure 1a shows the microbial community of the fermentation broth. In terms of bacteria, it was found that the dominant bacteria phyla were identified as *Proteobateria*, *Actinobateria*, *Firmicytes*, and *Bacteroidetes*. Twelve genera with high relative abundance (>1%) were identified as *Rhodobacter*, *Pseudomonas*, *Stenotrophomonas*, *Sphingobacterium*, *Enterococcus*, *Acinetobacter*, *Brevundimonas*, *Bifidobacterium*, *Klebsiella*, *Gemmobacter* and *Herminiimonas*, respectively. The microbes from the fermentation process should also play a vital role in the COD utilization and nitrogen removal. For instance, *Lactobacillus*, *Rhodobacter*, *Pseudomonas*, and *Sphingobacterium* were the dominant bacterial genera (>10%) in the fermentation broth, which were reported by the fermentation function with diverse substrates [28–30], partly contributing to the high COD utilization.

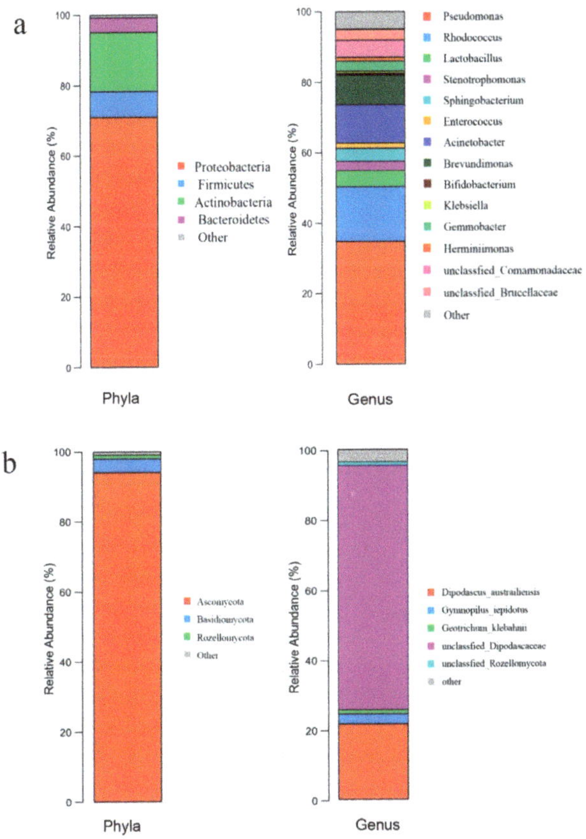

Figure 1. Relative abundance of dominant species in fermentation broth; (**a**,**b**) stands for the bacteria and fungi at phyla and genus level, respectively.

In terms of fungi, three phyla account for more than 94.9%, which were identified as *Ascomycota*, *Basidiomycota*, and *Rozellomycota*, respectively (Figure 1b), and five dominant species were identified as unclassified *Dipodascaceae* (61.76%), *Dipodascus australiensis* (21.67%), *Geotrichum klebahnii* (1.24%), *Gymnopilus lepidotus* (2.75%), and unclassified *Rozellomycota* (1.06%), respectively. These fungi were almost saprophytic, which could promote the degradation of cellulose and lignin in SMS.

3.2. Denitrification Performance and COD Removal

To investigate the effect of the fermentation broth on activated sludge formation and nitrogen removal ability, the SBR reactor was operated for about 30 days (Figures 2 and 3). The average influent NH_4^+-N concentrations of the control group and fermentation broth-added treatment were 88.20 and 80.95 mg/L, while the corresponding effluent concentrations were 19.74 and 20.63 mg/L, respectively. The average nitrogen removal rates between the control group (77.99%) and the fermentation broth-added treatment (74.44%) showed no significant difference (p = 0.155).

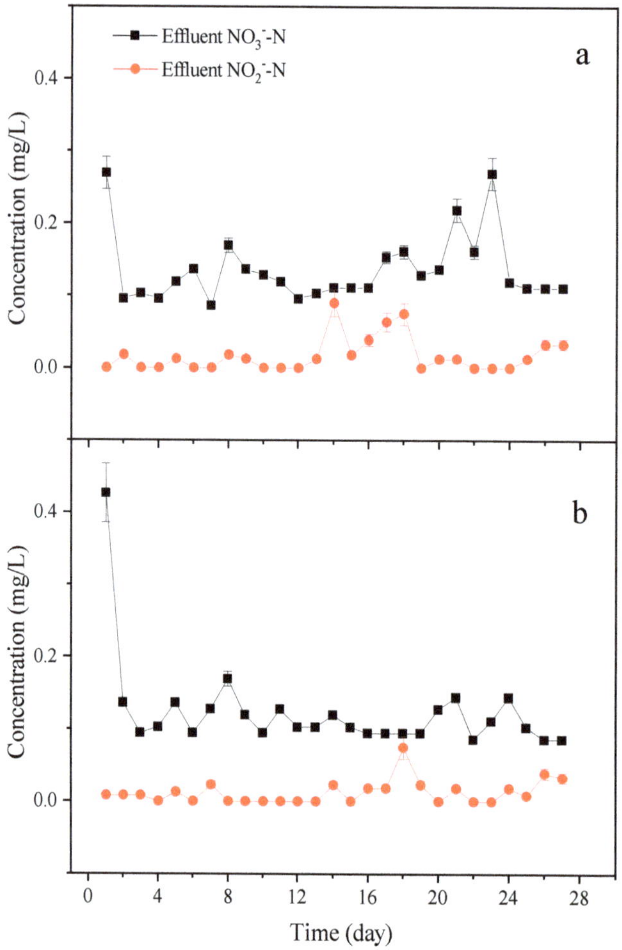

Figure 2. Nitrogen removal performance in the SBR reactors, (**a**) fermentation broth treated; (**b**) control group treated with glucose.

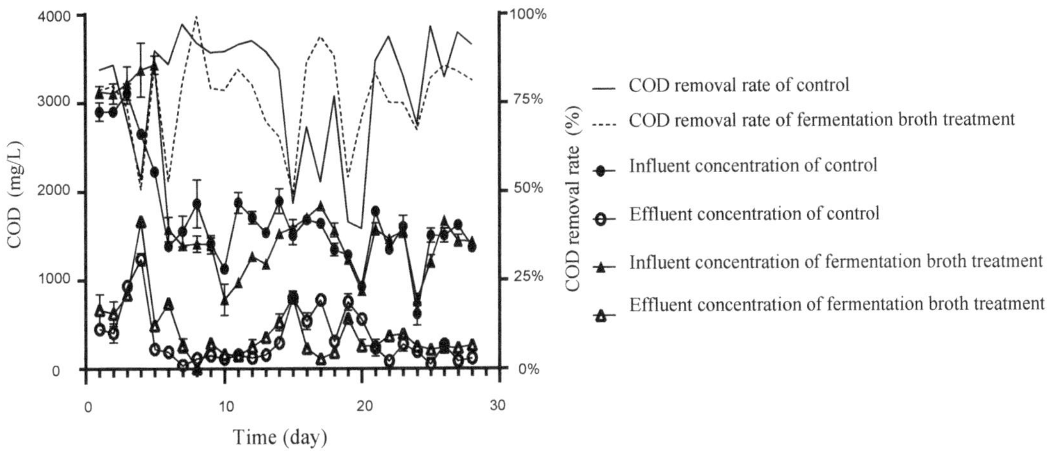

Figure 3. Variation of the COD concentration in two SBR systems.

Figure 2 shows the variation of NO_2^--N and NO_3^--N. Almost no accumulation of the two N species was found; the highest NO_2^--N concentration was lower than 0.09 mg/L in both of the two reactors. The maximum effluent NO_3^--N concentrations present on the first day were 0.426 and 0.269 mg/L in the control and fermentation broth-added treatments, respectively. As the operation proceeded, the NO_3^--N concentration in the control group was below 0.2 mg/L, which was lower than 0.269 mg/L in the treatment group. The low effluent NO_2^--N and NO_3^--N concentrations in the overall operation process also indicated the relatively high activity of nitrate reductase and nitrite reductase in the activated sludge.

The variation of the COD in the two reactors was also determined (Figure 3). In the first 5 days, a relatively high carbon source (C/N ratio = 13) was applied in order to promote the formation of flocculent-activated sludge. After that, the COD was adjusted to a normal level (about 1700 mg/L). The average effluent COD of the fermentation broth-added treatment (408.6 mg/L) was higher than the control group (344.6 mg/L), but the corresponding average COD removal rates showed no significant difference ($p = 0.3268$). Thus, it can be said the fermentation broth showed no negative effects on the COD and nitrogen removal of the SBR process.

3.3. Extracellular Polymeric Substance Composition

Hydrophilic PSs are an important part of a EPS, and their concentration variation in activated sludge was determined (Figure 4a). It can be found that PS concentrations in both reactors showed a similar increasing tendency to arrive at the first peak value on the 15th day. Fluctuation was observed after day 17 for the two groups, and the fermentation broth-added group presented its highest PS concentration (1288.77 mg/L) on day 21, while the PS in the control group ranged from 500 to 1000 mg/L. The average PS concentration in the fermentation broth-added treatment was 697.3 mg/L, which was significantly higher than the control group ($p = 0.0219$). The similar tendency of the PSs in the two reactors can be ascribed to both glucose and carbohydrate in the fermentation broth and favor PS synthesis [31,32].

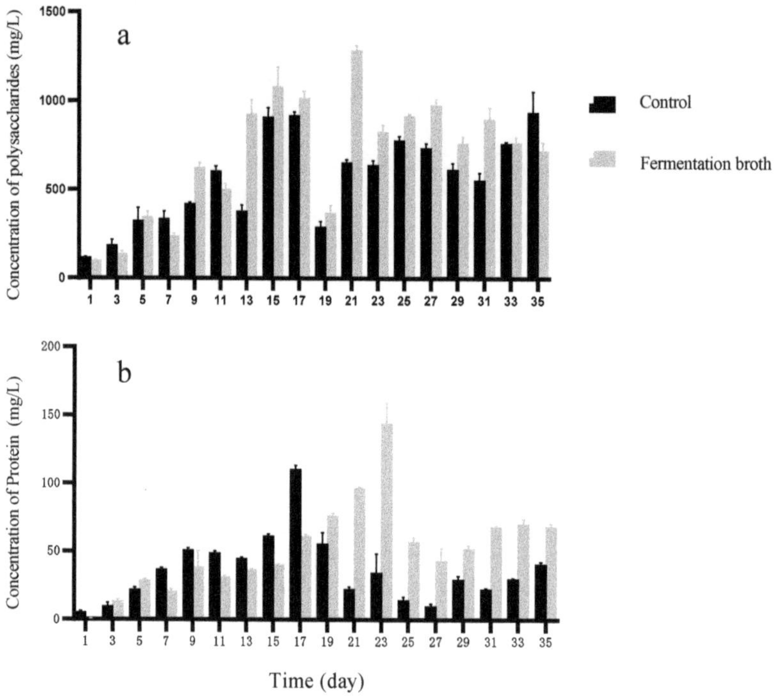

Figure 4. Concentration of PS (**a**) and PN (**b**) from activated sludge.

Figure 4b shows the variation of the PN concentration. Initially, PN in the control group increased as the operation proceeded and reached 110.90 mg/L on day 17, and then decreased to 50 mg/L after 5 days. In the treatment group, PN increased to a peak value (144.47 mg/L) on day 23, which was 30% higher than control. After the peak, PN in the fermentation broth-added reactor was maintained at a relatively high level, ranging from 50 to 60 mg/L. The higher PN concentration in the control group within the first 17 days may be ascribed to the easy degradation of glucose. It has been reported that an easily biodegradable organic carbon source leads to higher expression of biological activity [33]. The significantly higher PN concentration of the broth-added treatment in the further operation period may be ascribed to the gradually complete construction of the consortia. As a result, the abundant amino acids can be utilized to form PN, and the average PN concentrations were 52.59 mg/L in the treatment group, which was significantly higher than in the control (p = 0.0001).

It can be found that PS and NP reach a peak value and then start declining. This phenomenon can be ascribed to the influent COD concentration in both of the two groups decreased on day 17 and the starvation shock can reduce the bacterial metabolism, leading to the termination of the flocs bacteria in the sludge [34]. Thus, activated sludge became unstable and the EPS may present a relatively high degradable property. It is well known that EPS released by bacteria highly influenced the microbial aggregation, which were mainly determined by the carbon source here. According to our results, the addition of fermentation broth could promote the production of EPS, which is a benefit for the granulation of sludge. In a previous report, it was found that adding organic reject water increased the extracellular proteins/polysaccharides ratio of activated sludge, leading to higher adsorption and degradation of organic compounds [28]. Moreover, it has also been proven that over-production of EPS is a common self-protective strategy towards various external stresses. Thus, the higher EPS production in the fermentation-added treatment can be partly ascribed to the large amount of Cu^{2+} and Mg^{2+}. A higher EPS may also offer the activated sludge a better resistance ability toward environmental impact.

3.4. Absolute Abundance of Denitrification Functional Genes

Two typical nitrite reductases, *NirK* and *NirS*, were determined in activated sludge samples on day 10, which presented the most different beta diversity index compared with the original sample. As shown in Table 4, both the copies of *NirK* and *NirS* were lower than the control group, and the total nitrite reductase copies in the fermentation broth-added sample were only 25.27% of that in the control, indicating that adding the SMS fermentation broth showed a negative effect on the nitrogen removal function at an early stage. However, this negative effect can be removed since the whole nitrogen removal efficiency was comparable to the control group, which can be ascribed to the adaption of the microbial community.

Table 4. Nitrite reductase gene copies in day 10 samples.

Treatment	*NirS* (Average Copies)	*NirK* (Average Copies)
Control	26,316.472	15,039.141
Fermentation broth	6724.9544	3912.6366

3.5. Community Dynamics of Activated Sludge

Obviously, the nitrogen removal ability was recovered after domestication for 30 days, and investigation into the microbial structure dynamic could obtain a deep insight into the mechanism. Figure 5 presents the relative species abundance variation in the activated sludge at the phyla level. The predominant phyla were *Proteobacteria, Candidatus saccharibacteria, Bacteroidetes, Firmicutes, Actinobateria*, and *Verrucomicrobia*. Among the six phyla, *Proteobacteria, Candidatus saccharibacteria, Bacteroidetes, Firmicutes*, and *Actinobateria* were familiar bacteria in wastewater treatment plants with functions such as denitrification, dephosphorization, and biodegradation [35–38]. In terms of the control group, *Proteobateria* was dominant in the initial sample and sludge obtained from day 10. The abundance of *Candidatus saccharibacteria, Bacteroidetes*, and *Firmicutes* increased in the samples from both of the two reactors on day 10, while the *Proteobacteria* abundance decreased. It has been reported that *Candidatus saccharibacteria* prefers to enrich with complex carbon sources [39]. As domestication proceeded, *Candidatus saccharibacteria* showed obvious differences in the control and fermentation broth-added treatments and was much higher in the fermentation broth-added treatment on day 20. On day 30, *Candidatus saccharibacteria* became the predominant phylum in both two reactors with similar relative abundances. *Bacteroidetes* was the second priority phylum in the tested treatment, and was identified as *Proteobacteria* in the control group.

At the genus level (Figure 6), the predominant genus in the two reactors was identified as *unclassified Enterobacteriaceae* belonging to *Proteobacteria* initially. With the proceeding operation, the abundance of *unclassified Enterobacteriaceae* decreased, and the dominant genus was identified as *Saccharibacteria_ genera_incertae_sedis* at the end of domestication. This can be ascribed to the favorability of metabolizing various refractory pollutants, macromolecular organics, and complex carbon sources [40–42]. It was also found that the genera (>1%) can be divided as four clusters. The cluster composed of *Saccharibacteria_ genera_incertae_sedis, Flavobacterium*, unclassified *Rhodobacteraceae, Bdellovibrio*, and *Desulfovibrio, Trichococcus* occupied a dominant position during the domestication process. While the cluster of unclassified *Enterobacteriaceae, Acinetobacter, Rhizobium, Stenotrophomonas*, and *Comamonas* showed a decreased tendency during the operation. There were eight genera that presented higher abundances in the fermentation broth-added reactor rather than in the control group, identified as unclassified *Propionibacteriaceae*, unclassified *Bacteroidetes, Niabella*, unclassified *Sphingobacteriales*, unclassified *Flavobacteriaceae, Verrucomicrobium, Gemmobacter*, and *Runella*, respectively. These genera could highly contribute to the nitrogen removal ability of the sludge in treatment. For example, *Niabella* has been proven to oxidize NH_4^+-N [43], and *Bacteroidetes* can not only decompose complex carbon sources but also show nitrification ability [43,44].

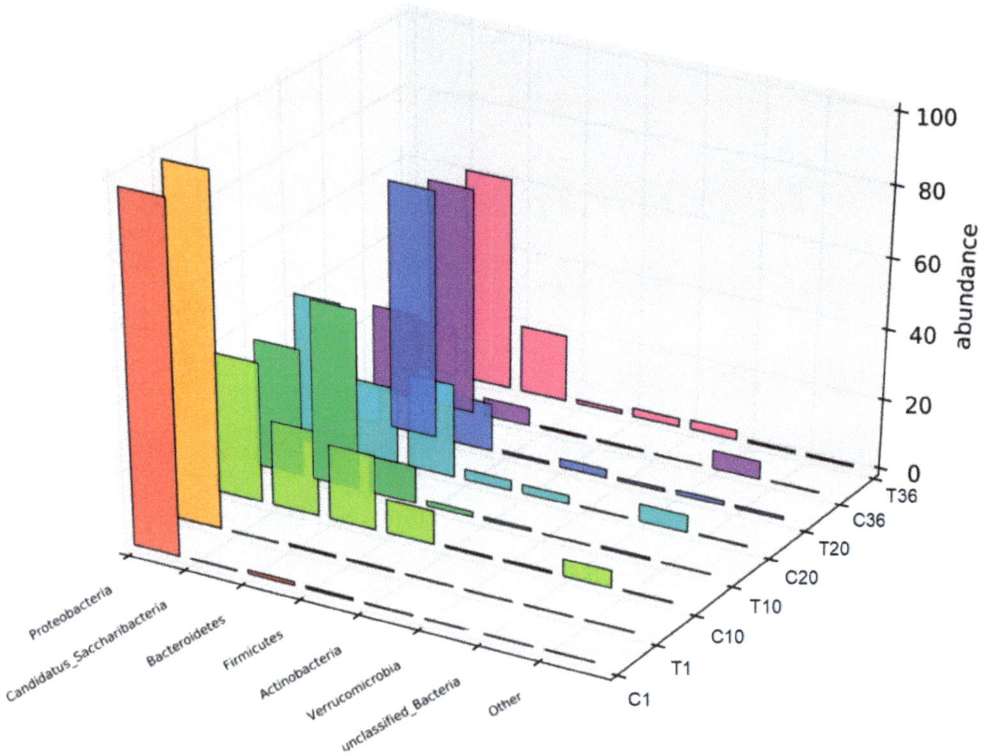

Figure 5. Bacterial diversity in activated sludge on a phyla level. C and T represent the control and broth-added treatments, respectively, and the number stands for the time (day) that the sample was obtained.

The dominant phyla of fungi were identified as *Ascomycota*, *Basidiomycota*, *Rozellomycota*, and unclassified phylum (Figure 7), and *Ascomycota* showed the highest abundance during the domestication process. These fungal phyla were widely identified in wastewater treatment plants [45]. The dynamic of the dominant genera (>1%) during the domestication is shown in Figure 8. Initially, *Aspergillus* occupied the dominant position in the activated sludge, which was replaced by *Dipodascus* and unclassified *Dipodascaceae* in the samples on day 10. It has been reported that *Dipodascaceae* can promote cellulose degradation in sawdust [46], which is beneficial for fermentation. The abundance of *Dipodascus* and unclassified *Dipodascaceae* decreased during the further domestication process, and the final fungal community was assembled with *Meyerozyma* and *Fusarium* as the dominant species. *Fusarium* species, such as *Fusarium solani*, was proven to show independent nitrification and denitrification abilities. It can be said that the fermentation broth showed a less obvious effect on the dominant fungal species.

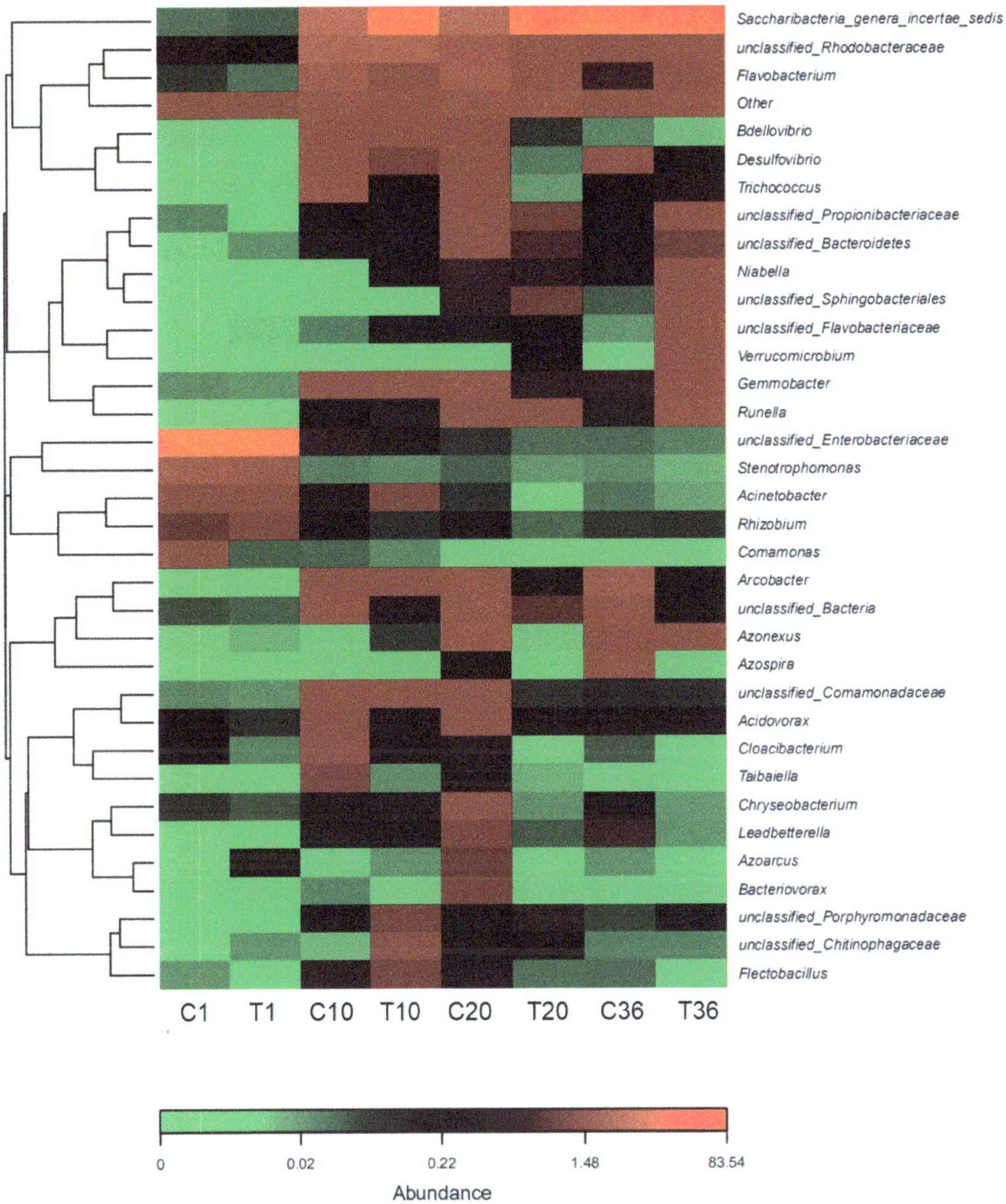

Figure 6. Bacterial diversity in the activated sludge on the genus level. C and T represent the control and broth-added treatment, respectively, and the number stands for the time (day) that the sample was obtained.

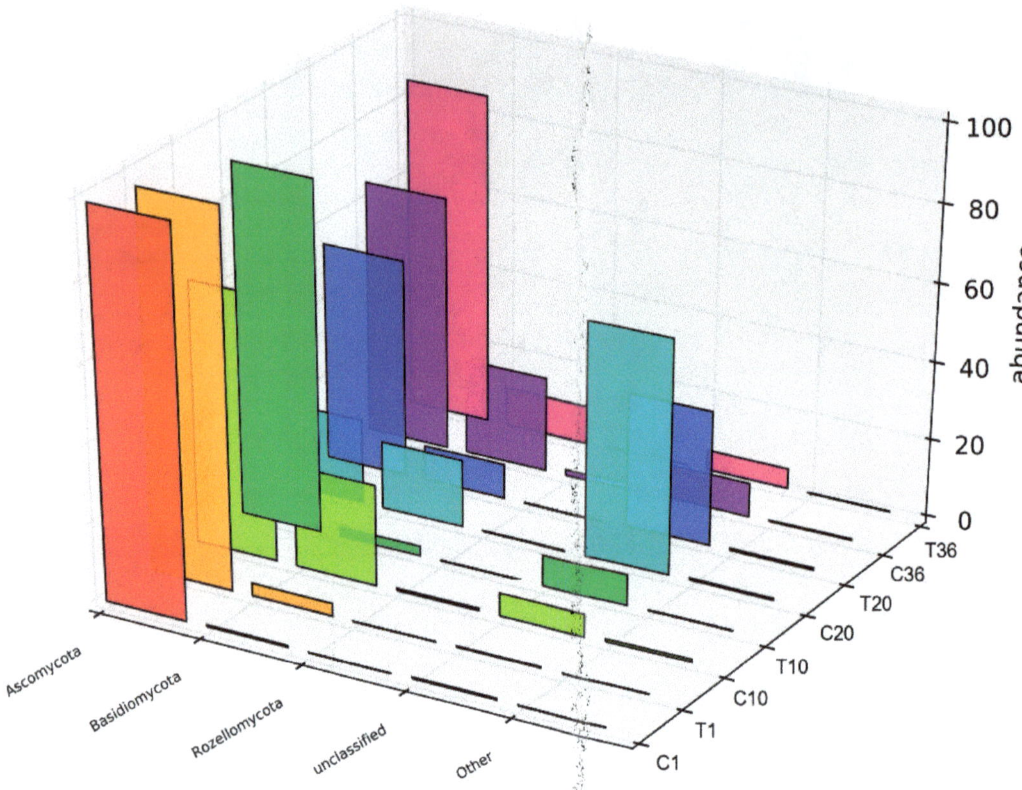

Figure 7. Fungal diversity in the activated sludge on the phyla level. C and T represent the control and broth-added treatment, respectively, and the number stands for the time (day) that the sample was obtained.

3.6. Environmental and Economic Benefits

It is estimated that each kilogram of fresh mushroom would produce approximately 5 kg of SMS [19]; thus, it is not only a challenge but also a huge fortune to recycle such amounts of waste. The economic benefit was evaluated according to the following rough calculation based on the COD concentration. The market price of SMS and glucose was about CNY 90.0 and CNY 1500/ton, respectively. To replace the COD from glucose (calculated as 1000 mg/L) with fermentation broth (20,000 mg/L), 12.5 kg of SMS was required for a 1000 L wastewater treatment. The required material cost was CNY 1.125 and CNY 1.5 for glucose. Besides the economic benefit, SMS recycling also avoids the environmental pollution caused by the incineration of SMS. Thus, using SMS to prepare the fermentation broth for wastewater treatment was eco-friendly and cost-effective.

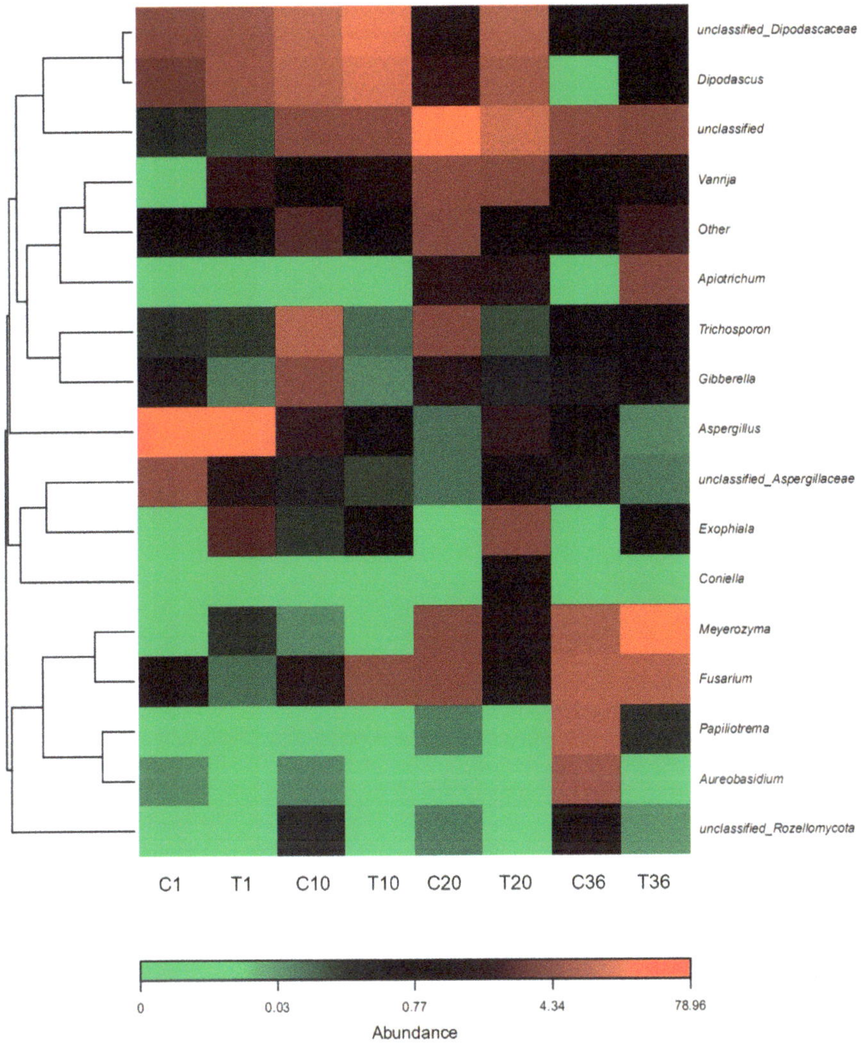

Figure 8. Fungal diversity in the activated sludge on the genus level. C and T represent the control and broth-added treatment, respectively, and the number stands for the time (day) that the sample was obtained.

4. Conclusions

In the present study, we employed SMS fermentation broth as an external carbon source for a nitrogen removal SBR system. The fermentation broth contains various nutrients such as proteins, polysaccharides, and organic acids, as well as mineral elements. In the fermentation broth, the dominant bacteria were *Rhodobacter*, *Lactobacillus*, and *Pseudomonas*, and the dominant fungi were *Saccharomycetales* and *Gymnopilus dilepis*. Compared with commercial external carbon sources, fermentation broth inhibited the nitrogen removal activity at the early stage, but a higher EPS formation and recovery phenomenon of nitrogen removal ability were observed after domestication for as long as 30 days. Microbial communities were highly influenced when 20% fermentation broth was added. As domestication proceeded, similar dominant species in the control and broth-feeding treatments were observed in both of the two groups. In general, SMS fermentation broth can be used as an external carbon source for nitrogen bioremoval.

Author Contributions: Data curation, W.Z. and X.B.; funding acquisition, R.C. and Y.Y.; investigation, W.Z. and X.B.; resources, Y.Y.; supervision, Y.Y.; writing-original draft, R.C.; writing-review and editing, Y.J. All authors have read and agreed to the published version of the manuscript.

Funding: This research was financially supported by the Major Program for Science and Technology Innovation of Wenzhou City (ZG2022038), the Public Research Project of Zhejiang Province (LQ23E090001), the Science and Technology Projects of Wenzhou City (S20220012), and the National Natural Science Foundation of China (42207433).

Data Availability Statement: The data presented in this study are available on request from the corresponding author.

Conflicts of Interest: The authors declare no conflict of interest.

Abbreviation

SMS	spent mushroom substrates
C/N	carbon–nitrogen ratio
EPS	extracellular polymeric substance
SBR	sequencing batch reactor
COD	chemical oxygen demand
qPCR	quantitative polymerase chain reaction
PN	extracellular protein
PS	extracellular polysaccharides
UV-vis	ultraviolet–visible spectrophotometer
HPLC	high-performance liquid chromatography
LC-MS	liquid chromatography mass spectrometry
ICP-OES	inductively coupled plasma emission spectrometry

References

1. Othman, N.Z.; Sarjuni, M.N.H.; Rosli, M.A.; Nadri, M.H.; Yeng, L.H.; Ying, O.P.; Sarmidi, M.R. Spent mushroom substrate as biofertilizer for agriculture application. In *Valorisation of Agro-Industrial Residues–Volume I: Biological Approaches*; Springer: Cham, Switzerland, 2020; Volume 3, pp. 7–57.
2. Foluke, A.; Olutayo, A.; Olufemi, A. Assessing spent mushroom substrate as a replacement to wheat bran in the diet of broilers. *Am. Int. J. Contemp. Res.* **2014**, *4*, 178–183.
3. Da Silva Alves, L.; de Almeida Moreira, B.R.; da Silva Viana, R.; Pardo-Gimenez, A.; Dias, E.S.; Noble, R.; Zied, D.C. Recycling spent mushroom substrate into fuel pellets for low-emission bioenergy producing systems. *J. Clean. Prod.* **2021**, *313*, 127875. [CrossRef]
4. Jin, Y.; Teng, C.; Yu, S.; Song, T.; Dong, L.; Liang, J.; Bai, X.; Liu, X.; Hu, X.; Qu, J. Batch and fixed-bed biosorption of Cd (II) from aqueous solution using immobilized Pleurotus ostreatus spent substrate. *Chemosphere* **2018**, *191*, 799–808. [CrossRef]
5. Liu, X.; Bai, X.; Dong, L.; Liang, J.; Jin, Y.; Wei, Y.; Li, Y.; Huang, S.; Qu, J. Composting enhances the removal of lead ions in aqueous solution by spent mushroom substrate: Biosorption and precipitation. *J. Clean. Prod.* **2018**, *200*, 1–11. [CrossRef]
6. Wu, J.; Zhang, T.; Chen, C.; Feng, L.; Su, X.; Zhou, L.; Chen, Y.; Xia, A.; Wang, X. Spent substrate of Ganodorma lucidum as a new bio-adsorbent for adsorption of three typical dyes. *Bioresour. Technol.* **2018**, *266*, 134–138. [CrossRef]
7. Liu, M.; Liu, X.; Wu, Z.; Zhang, Y.; Meng, Q.; Yan, L. Sulfur-modified Pleurotus ostreatus spent substrate biochar enhances the removal of cadmium in aqueous solution: Characterization, performance, mechanism. *J. Environ. Manag.* **2022**, *322*, 115900. [CrossRef]
8. Liu, F.; Tian, Y.; Ding, Y.; Li, Z. The use of fermentation liquid of wastewater primary sedimentation sludge as supplemental carbon source for denitrification based on enhanced anaerobic fermentation. *Bioresour. Technol.* **2016**, *219*, 6–13. [CrossRef] [PubMed]
9. Li, P.; Zuo, J.; Wang, Y.; Zhao, J.; Tang, L.; Li, Z. Tertiary nitrogen removal for municipal wastewater using a solid-phase denitrifying biofilter with polycaprolactone as the carbon source and filtration medium. *Water Res.* **2016**, *93*, 74–83. [CrossRef] [PubMed]
10. Choi, D.; Cho, S.; Jung, J. Key operating parameters affecting nitrogen removal rate in single-stage deammonification. *Chemosphere* **2018**, *207*, 357–364. [CrossRef]
11. Yu, G.; Peng, H.; Fu, Y.; Yan, X.; Du, C.; Chen, H. Enhanced nitrogen removal of low C/N wastewater in constructed wetlands with co-immobilizing solid carbon source and denitrifying bacteria. *Bioresour. Technol.* **2019**, *280*, 337–344. [CrossRef] [PubMed]

12. Feng, X.C.; Bao, X.; Che, L.; Wu, Q.L. Enhance biological nitrogen and phosphorus removal in wastewater treatment process by adding food waste fermentation liquid as external carbon source. *Biochem. Eng. J.* **2021**, *165*, 107811. [CrossRef]
13. Xiong, R.; Yu, X.; Yu, L.; Peng, Z.; Cheng, L.; Li, T.; Fan, P. Biological denitrification using polycaprolactone-peanut shell as slow-release carbon source treating drainage of municipal WWTP. *Chemosphere* **2019**, *235*, 434–439. [CrossRef]
14. Fu, X.; Hou, R.; Yang, P.; Qian, S.; Feng, Z.; Chen, Z.; Wang, F.; Yuan, R.; Chen, H.; Zhou, B. Application of external carbon source in heterotrophic denitrification of domestic sewage: A review. *Sci. Total Environ.* **2022**, *817*, 153061. [CrossRef] [PubMed]
15. Zhang, Z.; Zhang, Y.; Chen, Y. Recent advances in partial denitrification in biological nitrogen removal: From enrichment to application. *Bioresour. Technol.* **2020**, *298*, 122444. [CrossRef] [PubMed]
16. Xu, Z.; Dai, X.; Chai, X. Effect of different carbon sources on denitrification performance, microbial community structure and denitrification genes. *Sci. Total Environ.* **2018**, *634*, 195–204. [CrossRef] [PubMed]
17. Li, C.; Xu, S. Edible mushroom industry in China: Current state and perspectives. *Appl. Microbiol. Biotechnol.* **2022**, *106*, 3949–3955. [CrossRef]
18. Mohd Hanafi, F.H.; Rezania, S.; Mat Taib, S.; Md Din, M.F.; Yamauchi, M.; Sakamoto, M.; Hara, H.; Park, J.; Ebrahimi, S.S. Environmentally sustainable applications of agro-based spent mushroom substrate (SMS): An overview. *J. Mater. Cycles Waste Manag.* **2018**, *20*, 1383–1396. [CrossRef]
19. Moon, Y.H.; Shin, P.G.; Cho, S.J. Feeding value of spent mushroom (*Pleurotus eryngii*) substrate. *J. Mushroom* **2012**, *10*, 236–243.
20. Yang, Y.; Tao, X.; Lin, E.; Hu, K. Enhanced nitrogen removal with spent mushroom compost in a sequencing batch reactor. *Bioresour. Technol.* **2017**, *244*, 897–904. [CrossRef]
21. Hu, H.; Ma, S.; Zhang, X.; Ren, H. Characteristics of dissolved organic nitrogen in effluent from a biological nitrogen removal process using sludge alkaline fermentation liquid as an external carbon source. *Water Res.* **2020**, *176*, 115741. [CrossRef]
22. Tang, J.; Wang, X.C.; Hu, Y.; Pu, Y.; Huang, J.; Ngo, H.H.; Zeng, Y.; Li, Y. Nutrients removal performance and sludge properties using anaerobic fermentation slurry from food waste as an external carbon source for wastewater treatment. *Bioresour. Technol.* **2019**, *271*, 125–135. [CrossRef] [PubMed]
23. American Public Health Association. *Standard Methods for the Examination of Water and Wastewater*; American Public Health Association: Washington, DC, USA, 1985.
24. Jiang, F.; Feng, X.; Jiang, X.; Wang, P. Enhanced dewaterability of lake dredged sediments by electrochemical oxidation of peroxydisulfate on BDD anode. *Chemosphere* **2022**, *307*, 135832. [CrossRef]
25. Suzuki, M.; Hirai, T.; Arai, H.; Ishii, M.; Igarashi, Y. Purification, characterization, and gene cloning of thermophilic cytochrome cd1 nitrite reductase from Hydrogenobacter thermophilus TK-6. *J. Biosci. Bioeng.* **2006**, *101*, 391–397. [CrossRef] [PubMed]
26. Pomowski, A.; Zumft, W.G.; Kroneck, P.M.H.; Einsle, O. N_2O binding at a [4Cu: 2S] copper–sulphur cluster in nitrous oxide reductase. *Nature* **2011**, *477*, 234–237. [CrossRef] [PubMed]
27. Tocheva, E.I.; Rosell, F.I.; Mauk, A.G.; Murphy, M.E. Side-on copper-nitrosyl coordination by nitrite reductase. *Science* **2004**, *304*, 867–870. [CrossRef]
28. Wang, W.; Xie, H.; Wang, H.; Xue, H.; Wang, J.; Zhou, M.; Dai, X.; Wang, Y. Organic compounds evolution and sludge properties variation along partial nitritation and subsequent anammox processes treating reject water. *Water Res.* **2020**, *184*, 116197. [CrossRef]
29. Yu, D.; Feng, M.; Sun, J.; Xu, X.L.; Zhou, G.H. Protein degradation and peptide formation with antioxidant activity in pork protein extracts inoculated with Lactobacillus plantarum and Staphylococcus simulans. *Meat Sci.* **2020**, *160*, 107958. [CrossRef]
30. Wang, Y.; Fan, L.; Huang, J.; Liang, J.; Wang, X.; Ren, Y.; Li, H.; Yue, T.; Gao, Z. Evaluation of chemical composition, antioxidant activity, and gut microbiota associated with pumpkin juice fermented by Rhodobacter sphaeroides. *Food Chem.* **2023**, *401*, 134122. [CrossRef]
31. Kim, J.H.; Jang, Y.A.; Seong, S.B.; Jang, S.A.; Hong, S.H.; Song, J.K.; Eom, G.T. High-level production and high-yield recovery of lactobionic acid by the control of pH and temperature in fermentation of Pseudomonas taetrolens. *Bioprocess Biosyst. Eng.* **2020**, *43*, 937–944. [CrossRef]
32. Shankar, K.; Kulkarni, N.S.; Jayalakshmi, S.K.; Sreeramulu, K. Saccharification of the pretreated husks of corn, peanut and coffee cherry by the lignocellulolytic enzymes secreted by Sphingobacterium sp. ksn for the production of bioethanol. *Biomass Bioenergy* **2019**, *127*, 105298. [CrossRef]
33. Geyik, A.G.; Kılıç, B.; Çeçen, F. Extracellular polymeric substances (EPS) and surface properties of activated sludges: Effect of organic carbon sources. *Environ. Sci. Pollut. Res.* **2016**, *23*, 1653–1663. [CrossRef]
34. Sponza, D.T. Extracellular polymer substances and physicochemical properties of flocs in steady and unsteady-state activated sludge systems. *Process Biochem.* **2002**, *37*, 983–998. [CrossRef]
35. Miao, L.; Zhang, Q.; Wang, S.; Li, B.; Wang, Z.; Zhang, S.; Zhang, M.; Peng, Y. Characterization of EPS compositions and microbial community in an Anammox SBBR system treating landfill leachate. *Bioresour. Technol.* **2018**, *249*, 108–116. [CrossRef] [PubMed]
36. Li, X.Y.; Yang, S.F. Influence of loosely bound extracellular polymeric substances (EPS) on the flocculation, sedimentation and dewaterability of activated sludge. *Water Res.* **2007**, *41*, 1022–1030. [CrossRef]
37. Su, F.; Wang, Z.; Huang, T.; Zhang, H.; Zhang, H. Simultaneous removal of nitrate, phosphorous and cadmium using a novel multifunctional biomaterial immobilized aerobic strain *Proteobacteria Cupriavidus* H29. *Bioresour. Technol.* **2020**, *307*, 123196. [CrossRef]

38. Li, X.; Lu, Y.; Luo, H.; Liu, G.; Zhang, R. Microbial stratification structure within cathodic biofilm of the microbial fuel cell using the freezing microtome method. *Bioresour. Technol.* **2017**, *241*, 384–390. [CrossRef]
39. Hosseinzadeh, A.; Zhou, J.L.; Navidpour, A.H.; Altaee, A. Progress in osmotic membrane bioreactors research: Contaminant removal, microbial community and bioenergy production in wastewater. *Bioresour. Technol.* **2021**, *330*, 124998. [CrossRef]
40. Ma, X.; Wang, X.; Liu, Y.; Gao, J.; Wang, Y. Variations in toxicity of semi-coking wastewater treatment processes and their toxicity prediction. *Ecotoxicol. Environ. Saf.* **2017**, *138*, 163–169. [CrossRef] [PubMed]
41. Hanada, A.; Kurogi, T.; Giang, N.M.; Yamada, T.; Kamimoto, Y.; Kiso, Y.; Hiraishi, A. Bacteria of the candidate phylum TM7 are prevalent in acidophilic nitrifying sequencing-batch reactors. *Microbes Environ.* **2014**, *29*, 353–362. [CrossRef]
42. Zhao, J.; Li, Y.; Chen, X.; Li, Y. Effects of carbon sources on sludge performance and microbial community for 4-chlorophenol wastewater treatment in sequencing batch reactors. *Bioresour. Technol.* **2018**, *255*, 22–28. [CrossRef]
43. Xing, W.; Li, D.; Li, J.; Hu, Q.; Deng, S. Nitrate removal and microbial analysis by combined micro-electrolysis and autotrophic denitrification. *Bioresour. Technol.* **2016**, *211*, 240–247. [CrossRef] [PubMed]
44. Wang, X.; Xing, D.; Ren, N. p-Nitrophenol degradation and microbial community structure in a biocathode bioelectrochemical system. *RSC Adv.* **2016**, *6*, 89821–89826. [CrossRef]
45. Assress, H.A.; Selvarajan, R.; Nyoni, H.; Ntushelo, K.; Mamba, B.B.; Msagati, T.A. Diversity, co-occurrence and implications of fungal communities in wastewater treatment plants. *Sci. Rep.* **2019**, *9*, 14056. [CrossRef] [PubMed]
46. Wang, K.; Ma, X.; Yin, X.; Wu, C.; Wang, Z.; Wu, Y.; Zhao, Y.; Tian, Y. Difference and interplay of microbial communities, metabolic functions, trophic modes and influence factors between sludge and bulking agent in a composting matrix. *Bioresour. Technol.* **2021**, *336*, 125085. [CrossRef] [PubMed]

water

MDPI

Article

Genotoxic Effects on *Daphnia magna* Fed with Aquatic Green Algae Exposed to Silver Nanoclusters

Li Zhang [1,*] and Haoqiang Tan [2]

[1] Department of Environmental Engineering, School of Life Sciences, Taizhou University, Taizhou 318000, China
[2] School of Civil Engineering and Architecture, Taizhou University, Taizhou 318000, China; tanhaoqiang126@126.com
* Correspondence: zhangli8157@sina.com

Abstract: Ag nanoclusters (AgNCs) have gained widespread applications in recent years due to their excellent antimicrobial efficacy and distinctive molecule-like characteristics. However, concerns about their potential effects on environmental and human health have been raised. Despite the fact that abundant research has been carried out to examine the possible ecotoxicology of AgNCs in a variety of living organisms, these studies have mostly concentrated on the toxicology of individual organisms and only a few have attempted to look into the impact of AgNCs across the aquatic food chain. This work evaluated the transcriptome level genotoxicity of AgNCs and their degraded Ag ions in two model species food chains: the aquatic green algae *Scenedesmus obliquus* and the invertebrate *Daphnia magna*. *Daphnia magna*'s digestive system and glycerophospholipid metabolism were hindered after feeding on Ag-containing algae as a result of down-regulation of the crucial gene PLA2G(SPLA2) that codes for secretory phospholipase A2. Our research also showed that the genotoxicity of AgNCs to *Daphnia magna* was mediated by a synergic interaction between the particulate form of AgNCs and their degraded Ag ions. The current work offers a fresh viewpoint on the mechanisms underlying AgNCs' harmful effects and the possible ecological concern that metal-based nanoparticles provide to aquatic life.

Keywords: Ag nanoclusters; *Scenedesmus obliquus*; *Daphnia magna*; food chain; transcriptome

1. Introduction

When compared to their larger counterparts, silver nanoparticles (AgNPs, >2 nm), silver nanoclusters (denoted AgNCs) with particle sizes below 2 nm, typically composed of several to one hundred silver atoms, exhibit obviously different physical and chemical properties [1–3]. Due to the significant quantum size confinement of free electrons in this sub-2 nm size range, AgNCs possess discrete electronic states and display distinctive molecule-like properties, such as intensive luminescence, well-defined molecular structure, quantized charge and HOMO–LUMO transitions [2,4–8]. In particular, the unique properties of AgNCs coupled with their ultra-small size and good biocompatibility have enabled them to emerge as promising functional materials for a wide variety of biomedical applications, especially as effective antimicrobial agents and as optical probes for bioimaging and biolabeling applications [2,6,9–12]. AgNCs will unavoidably be released into the environment as a result of their extensive use, especially in the aquatic environment. Given the growing public concern over the possible threats they pose to human health and living things, it is therefore worthwhile to investigate the potential ecotoxicology of AgNCs.

Although numerous studies have been conducted to explore the potential ecotoxicology of large AgNPs in a variety of microorganisms [12,13], algae [14], plants [15], invertebrates [16,17] and vertebrates [18,19], these studies have largely concentrated on the effect on individual organisms, and only a few have described the toxicology of ultra-small AgNCs. The trophic toxicology of AgNCs to the aquatic food chain is a significant issue that warrants further research because several previous studies have shown that nanoparticles, including AgNPs, could potentially accumulate in aquatic ecosystems and eventually cause harm to the "top-level" species [20–22]. As is generally known, freshwater algae, a great example of a unicellular organism, play a significant role in both photosynthesis and the cycle of carbon dioxide on a global scale [23,24]. *Daphnia magna* (*D. magna*) represents a model zooplankton in the aquatic environment with which to illuminate the toxic mechanisms of AgNPs in environmental assessments [17,25]. According to a previous study [22], freshwater algae have the capacity to internalize AgNPs, which may then be transmitted to higher tropic-level organisms like *Daphnia magna*. Therefore, we believe that through the food chain, algal-borne AgNCs may ultimately be hazardous to *Daphnia magna*.

In the past few years, many acute toxicity tests (short-term exposure and at high Ag concentrations) have been used to investigate the effects of AgNPs on freshwater organisms; however, environmental exposures to AgNPs are more likely to be long-term and at low concentrations [26–29]. In other words, testing for chronic toxicity is more relevant than testing for acute toxicity when evaluating the environmental consequences of AgNPs. Environmental conditions are also dynamic and complicated, and they are not easily reversible [26,27]. As a result, it is challenging to conduct an experiment in the lab to examine the impacts of AgNPs on the environment. If research on the long-term impacts of exposure to AgNPs could be done at the molecular level (for example, utilizing transcriptomic approaches), then the results should be able to provide useful information [30]. The biomolecular impacts of AgNPs and the released Ag ions on aquatic organisms have not received much attention until now. Transcriptomic analysis has recently been used to study, in great detail, small variations in gene expression in aquatic organisms after exposure to AgNPs. These investigations have provided the foundation for a better understanding of the molecular mechanisms behind the toxicity caused by AgNCs [14,31,32].

The goal of the current work is to use cutting-edge transcriptome technology to examine the molecular harmful effects of AgNCs carried by algae on the aquatic invertebrate *Daphnia magna*. Our results showed that *Daphnia magna* exposed to AgNCs carried by algae showed blatant changes in gene expression across the food chain. We suggest that the current study offers a fresh viewpoint on how to identify AgNCs-mediated toxicity as well as an early warning system for environmental risk brought on by the accumulation of other metal nanoparticles in aquatic life.

2. Materials and Methods

2.1. Materials

Prior to all experiments, all glassware and pipet tips were soaked in 0.05 M nitric acid for 24 h and then rinsed with Milli-Q water (18.2 MΩ). Silver nitrate (99.9999%), sodium borohydride ($NaBH_4$), L-glutathione reduced (GSH) and L-cysteine ($\geq$99%) were purchased from Sigma-Aldrich. The other reagents (analytical or ultrapure grades) were purchased from Sinopharm Chemical Reagent Co., Ltd. (Shanghai, China). All solutions used in this study were prepared in Milli-Q water. Ultra-small AgNCs were prepared according to the method of Yuan [9,33]. The AgNC solutions were subsequently stored at 4 °C for later use. Further details on the characterization of AgNCs and their dissolution in exposure water are available in our previous work [34,35].

2.2. Algae Culture and Nanosilver Exposure Conditions

Scenedesmus obliquus (No. FACHB-417), was obtained from the Institute of Hydrobiology of the Chinese Academy of Sciences (Wuhan, China) and subsequently cultured in SE medium (Wuhan, China, Supplementary Materials Table S1) under controlled conditions (12:12 h light/dark photoperiod; 25 °C; illumination 2500~3000 lx; artificial shaking three times every day) in SE. To avoid biological contaminations, all glassware and algae culture media were sterilized in a SANYO sterilizing oven (Osaka, Japan) (121 °C, 30 min).

For algae exposure experiments, in consideration of the concentration level of AgNPs (0.1~146 μg/L) in the aquatic environment [28], algae in the log-phase of growth

(5~10 × 10^5 cells/mL) were exposed to AgNC concentrations (according to the concentration of Ag atoms) of 0.00, 135 and 135 μg/L (containing 0.5 mM L-cysteine) and silver ions (5.0, 10.0 and 20.0 μg/L), respectively. The addition of L-cysteine, at a concentration of 0.5 mM, was sufficient to inhibit the dissolution of Ag^+ into SE medium. According to our previous study, the concentration level of Ag^+ released from AgNCs (μg/L) is almost equal to that of silver ions (5.0, 10.0 and 20.0 μg/L) dissolved from silver nitrate [34]. The control experiments were exposed to SE medium (without silver). The exposure experiments were carried out in triplicate.

2.3. Daphnia Magna Culture and Ag Exposure

Adult *Daphnia magna* were obtained commercially (Institute of Hydrobiology of Chinese Academy of Sciences) and grown in standard dilution water medium (SDW medium: 294 mg/L $CaCl_2{\cdot}2H_2O$, 123 mg/L $MgSO_4{\cdot}7H_2O$, 65 mg/L $NaHCO_3$, 6.3 mg/L KCl). The growth conditions were the same as those of *Scenedesmus obliquus* mentioned above. Prior to the silver exposure experiments, the *Daphnia magna* were fed with SE medium-grown *Scenedesmus obliquus* (containing no AgNCs or $AgNO_3$).

After exposure to Ag for 96 h, *Scenedesmus obliquus* was harvested on a 0.45 μm filter membrane for feeding to *Daphnia magna*. The algae cells were washed with Milli-Q water three times to remove remanent AgNCs or $AgNO_3$ on the cell surface. The cells were then resuspended in SDW medium to reach a cell density of ~5 × 10^5 cells/mL. Finally, 100 mL of SDW medium containing algae cells (~5 × 10^5 cells/mL) was introduced into eighteen beakers (Figure S1, one control and five AgNCs or $AgNO_3$ treatments). Algae cells were quickly consumed by *Daphnia magna* within 30 min. Considering the extremely small volume of *Daphnia magna*, each beaker contained ~100 *Daphnia magna* in order to ensure adequate RNA samples. The volume of SDW medium in each breaker remained relatively constant (1000 mL) through the addition of fresh medium. The experiment was done in triplicate.

2.4. RNA Sequencing

According to our previous studies using confocal microscopic images and TEM images, AgNCs are indeed internalized by the algae cells [34,35]. At the end of 48 h of exposure to algae cells grown with or without AgNCs/$AgNO_3$, *Daphnia magna* fed with green algae but without silver were used as controls (sample A). *Daphnia magna* fed with green algae treated with 135 μg/L r-AgNCs but without 0.5 mM L-cysteine (sample B), and fed with green algae treated with the 135 μg/L r-AgNCs (with 0.5 mM L-cysteine) (sample C) and those fed with green algae treated with 10 μg/L silver ions (sample D) were collected and washed three times with Milli-Q water. To obtain sufficient RNA, the samples of each treatment were mixed and then immediately frozen in liquid nitrogen. Total RNA extraction of *Daphnia magna* was performed according to the manufacturer's instructions for trizol reagent (Invitrogen, Carlsbad, CA, USA). The concentration and integrity of RNA was detected by a Nanodrop 2000 Spectrophotometer and a 2100 Bioanalyser (Agilent, Palo Alto, CA, USA), respectively. After the cDNA library of each treatment was prepared by polymerase chain reaction, the sequencing libraries were gradually diluted and quantified to 4–5 pM and sequenced on the Illumina NextSeq 500 platform.

2.5. Transcriptomic Analysis

A pair-end sequencing strategy was employed to generate RNA-Seq raw reads with an average length of 150 bp. In order to better conduct de novo transcriptome assembly, the raw reads were filtered to obtain high-quality reads by removing the adapter sequences and low-quality bases at the 3′ end (Q < 20) using the FastQC program(version: 0.11.2). Quality reads were converted into contigs, established de Bruijn graphs and optimized de Bruijn graphs to obtain the final transcripts, using the Trinity programme (http://trinityrnaseq.sf.net) (version: r20140717, accessed on 10 July 2023). The transcripts were subsequently subjected to the NCBI basic local alignment search tool (BLAST) (http://

www.ncbi.nlm.nih.gov/) (version: 2.2.30+, accessed on 15 July 2023) against the NCBI non-redundant protein database (NR) (http://www.ncbi.nlm.nih.gov/) (version: GRCh36, accessed on 12 July 2023) (cut-off E-value < 10^{-5}). The top-hit transcripts were identified as unigenes. For functional annotation of all the unigenes, we searched them against the public databases using NCBI BLAST, including the NCBI non-redundant protein database, the Gene Ontology (GO) database (http://geneontology.org/) (version: Conesa A and Gotz, accessed on 15 July 2023), Kyoto Encyclopedia of Genes and Genomes (KEGG) database (http://www.genome.jp/kegg) (version: 81.0, accessed on 10 July 2023) and the Evolutionary genealogy of genes: non-supervised Orthologous Groups (eggNOG) database (http://eggnog.embl.de/) (version: 4.0.beta/, accessed on 12 July 2023).

To analyse the differentially expressed unigenes derived from *Daphnia magna* after exposure to algal-borne AgNCs and Ag NO_3 compared to the controls, the number of reads per kilobase of the exon model per million mapped reads value (RPKM) was used to normalize the gene expression at the transcriptional level. Differentially expressed unigenes were identified by using the DESeq program (http://www-huber.embl.de/users/anders/DESeq) (version: 1.18.0, accessed on 10 July 2023). The false discovery rate (FRD, <0.05) method was introduced to determine the threshold of the *p*-value in multiple tests and further judge the significance of the difference in gene expression. When greater than a two-fold change (absolute value of $\log_2$ fold change > 1) and *p*-value < 0.05 in gene expression were observed in our study, unigenes were considered to have significant differential expression. KEGG Orthology (KO) metabolic pathways analysis of the differentially expressed unigenes was carried out on the public website (http://www.genome.jp/kegg/tool/map_pathway2.html) (version: 2015/05/22, accessed on 18 July 2023). GO and KEGG enrichment analysis of differentially expressed unigenes were conducted using the Goatools software (https://github.com/tanghaibao/GOatools) (version: v1.2.3, accessed on 10 July 2023) and the KOBAS software (http://kobas.cbi.pku.edu.cn/home.do) (version: 2.0, accessed on 15 July 2023), respectively. Cluster analysis of differential unigenes was performed by means of Cluster3.0/TreeView programme (http://bonsai.hgc.jp/~mdehoon/software/cluster/manual/index.html) (version: v1.2.0, accessed on 10 July 2023).

3. Results and Discussion

3.1. Sequencing and De Novo Transcriptome Assembly

Before cDNA libraries were constructed in our study, the concentration and integrity of total RNA were firstly examined by a UV-vis spectrophotometer (Nanodrop 2000, Waltham, MA, USA) and agarose gel electrophoresis, respectively (Table S2 and Figure S2). To obtain the *Daphnia magna* transcriptome expression profile after exposure to algal-borne AgNCs and Ag NO_3, four independent cDNA libraries from various *Daphnia magna* treatments (samples A, B, C and D) were sequenced using the Illumina NextSeq 500 platform. After sequencing and data filtering, the raw reads and high-quality reads of various samples are presented in Table S3. Subsequently, transcript de novo assembly for quality reads was carried out by Trinity software(version: r20140717). A summary of all contigs, transcripts and unigene assemblies is shown in Table S3. A total of 32,009 unigenes were obtained with a maximum length of 237,122 bp, an average length of 923 bp and an N50 of 1517 bp, and these unigenes were further used to perform the analysis of differentially expressed genes at the transcriptional level. The length distribution of unigenes is displayed in Figure S3. The lengths of unigenes were ranged from 201 bp to 23,712 bp, and the most abundant unigenes were clustered in a group 200–299 bp in length. For further functional prediction and classifications, all unigenes were searched against several public databases, including the NCBI NR database, the GO database, eggNOG database and KEGG database, among others. The results of the annotation of the *Daphnia magna* transcriptome are summarized in Table S4, Figures S4–S6. Among the 32,009 unigenes, 32,009, 29,260 and 26,943 had significant hits in the NR, eggNOG and SwissProt databases, respectively.

3.2. Differential Expression Gene Analysis

A statistical analysis of differentially expressed genes between the three silver treatments and the controls was conducted using the gene expression data. Volcano plots showed the distribution trends of differential expression genes (red spots) and genes with no differential expression (blue spots) (Figure S7). It is quite clear that the expression levels of many differentially expressed genes derived from *Daphnia magna* after feeding with green algae treated with three silver (r-AgNCs, r-AgNCs containing L-cysteine and silver ions) were up- and down-regulated compared to the control treatment, which further demonstrates that r-AgNCs and their dissolved silver ions cause a toxic effect on the gene expression of *Daphnia magna*. The number of up- and down-regulated differentially expressed genes obtained per treatment group was coincident with the overlap between the differentially expressed genes for each silver treatment (Figure 1a,b). Figure 1c illustrates the heatmap of differentially expressed genes between the various algal-borne AgNC treatments and the control. The gene expression levels are visualised by using a plot colour ranging from green (low expression) to red (high expression), which demonstrates the significant differences in the level of gene expression for differentially expressed genes between each algal-borne AgNCs treatment and the control. The condition tree also shows that the gene expression profiles for differentially expressed genes were consistent across particle-specific AgNCs and their dissolved silver ions.

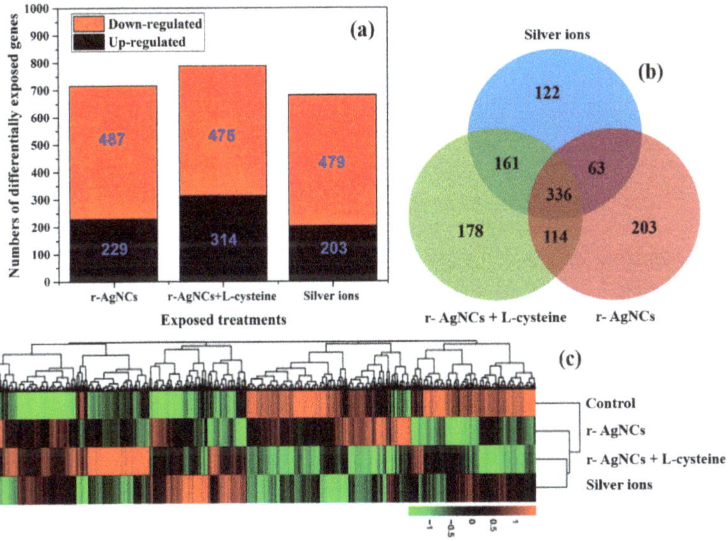

Figure 1. Differentially expressed genes of *Daphnia magna* resulting from a diet of green algae exposed to three silver treatments (r-AgNCs, r-AgNCs + L-cysteine and silver ions) compared to the controls (*Daphnia magna* on a diet of green algae only). (**a**) Colour bars indicate the number of (red) up- and (blue) down-regulated genes derived from *Daphnia magna* (|fold change| > 2 and p-value < 0.05). (**b**) Venn diagram showing numbers of differentially expressed genes from *Daphnia magna* after the diet of green algae (|fold change| > 2 and p-value < 0.05). (**c**) Heatmap of differentially expressed genes (|fold change| > 2 and p-value < 0.05) derived from *Daphnia magna* as a result of the diet of green algae. Colour bars indicate the gene expression levels: (red) up-regulated genes and (green) down-regulated genes.

3.3. Enrichment Analysis of the Differential Expression Genes

The biological significance of the changes in gene expression of differentially expressed genes in *Daphnia magna* fed on green algae treated with silver has been investigated by using enrichment analysis in GO terms (denoted as GOSlim in our study, Figure S8) and the KEGG metabolic pathway (Figure 2). Gene Ontology (GO) is a standardized gene functional classification system including the terms for biological processes, cellular components and molecular functions. The KEGG metabolic pathway, which represents our knowledge of the molecular interaction and reaction networks of *Daphnia magna*, can help further our understanding of the biological functions of differentially expressed genes and how these genes interact. Enrichment analysis of differentially expressed genes (up- and down-regulated unigenes) was conducted to gain an overall perspective of the biological effects of each exposure to the green algae diet. The most significantly typical GO terms (structural molecule activity and external encapsulating structure) and KEGG pathways (lipid metabolism and digestive system) remained common across all three different algae diets, suggesting that there was a very significant overlap in the gene expression of differentially expressed unigenes in response to the three kinds of silver treatments. Based on this overlap, it is likely that the toxicity of AgNCs to *Daphnia magna* was predominantly due to the combined action between the AgNCs and their released silver ions.

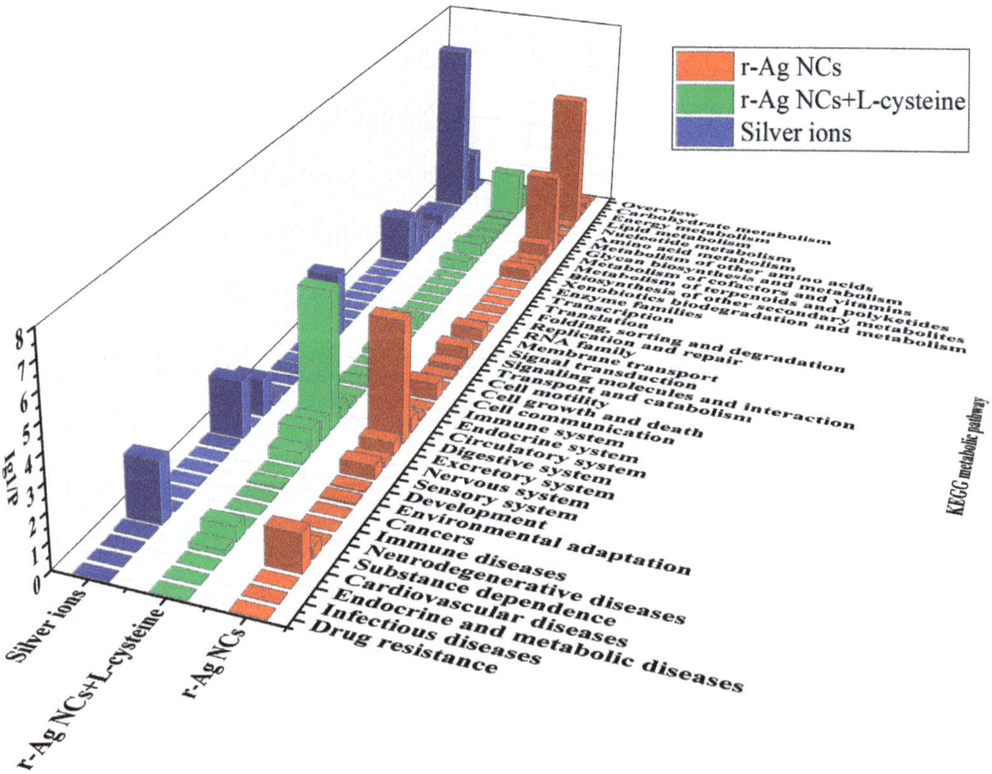

Figure 2. KEGG enrichment analysis of differentially expressed genes of *Daphnia magna* from the diet of green algae treated with silver (AgNCs, AgNCs with L-cysteine and Ag ions). The enrichment analysis was conducted between silver treatments and controls (p values < 0.05) after 48 h of exposure, using all expressed unigenes as a background.

Scenedesmus obliquus, a freshwater green algae, is a well-known primary producer in the aquatic ecosystem and a source of food for primary or higher trophic-level consumers, including the freshwater cladoceran *Daphnia magna*. As is shown in Figure 2, the enrichment of differentially expressed unigenes in the KEGG metabolic pathway (lipid metabolism and digestive system) was more significant than other metabolic pathways (p-value < 0.05), which demonstrates that the processes of metabolism of *Daphnia magna* were inhibited at the molecular level when fed with the three different diets of green algae. Furthermore, Ig(1/p) of the terms of external encapsulating structure and structural molecular activity were larger than that of other terms of GO enrichment (Figure S8, adjusted $p < 0.05$), indicating that the differentially expressed genes coding the cellular components (e.g., cell membrane) and the biological macromolecules in the body (e.g., metabolic enzymes) were regulated after exposure to algal-borne AgNCs and Ag ions, and it is anticipated that this regulation was also associated with the process of metabolism in *Daphnia magna*. As a result, the AgNC-produced Ag ions and the diets of green algae that the *Daphnia* were treated with had a greater effect on their metabolism.

3.4. KO Metabolic Pathway Analysis of the Differential Expression Genes

As shown in Figure S1, after *Daphnia* being fed the diet of algae treated with AgNCs and Ag ions for 48 h, the content of *Scenedesmus obliquus* in the beaker decreased obviously, indicating that *Daphnia magna* can feed on *Scenedesmus obliquus*. Therefore, it is essential to investigate the effects of the diet of algae on the digestive system of *Daphnia magna*, especially with respect to the subtle differences in gene expression. Differentially expressed genes in the digestive system of *Daphnia magna* after being fed a diet of algae treated with silver for 48 h are listed in Table S5. The gene expression analysis demonstrated that many of the unigenes encoding for key enzymes and protein subunits of the digestive system were up- and down-regulated compared to the controls. Among all these genes, PLA2G and SPLA2 (gene ID: c75537_g1_i1) were especially sensitive to AgNC and Ag^+ ion exposure. These genes encode for secretory phospholipase A2, a key enzyme for fat digestion and absorption, and they were down-regulated at the transcriptome level. Interestingly, PLA2G and SPLA2 (gene ID: c22033_g1_i2) also encode for secretory phospholipase A2 and they were also down-regulated as a result of exposure to AgNCs and Ag^+ ions, except for AgNCs + L-cysteine, which suggests that a protein, such as secretory phospholipase A2, can be coded by different genes. Since the digestive system of *Daphnia magna* relies heavily on the lipase secreted by the hepatopancreas (Figure 3a), a down-regulation of the genes PLA2G and SPLA2, which encode phospholipase A2, suggests that a diet of algae treated with AgNCs and the subsequently released Ag ions may have an impact on how fats (such as glycerophospholipid) are metabolised.

Apart from the effect on the digestive system of *Daphnia magna*, the diet of algae treated with AgNCs and their released Ag ions can also lead to changes in glycerophospholipid metabolism and showed similarly significant alterations in the expression levels of differentially expressed genes (Table S5, gene ID: c75537_g1_i1). Glycerophospholipid is one of the most important membrane lipids and can be hydrolyzed to produce aliphatic acid and glycerol by the catalysis of secretory phospholipase A2 (Figure 3b). Aliphatic acid will further produce acetyl coenzyme A by β-oxidation and finally enter the tricarboxylic acid cycle (TCA cycle). Furthermore, intermediate products of β-oxidation (e.g., arachidonic acid metabolism, linoleic acid and alpha-linolenic acid) were also regulated by secretory phospholipase A2 (Table S5). The down-regulation of key enzymes, such as secretory phospholipase A2 coded by gene PLA2G (SPLA2), indicates that glycerophospholipid metabolism was probably also regulated as a result of the diet of algae treated with AgNCs and their released Ag ions, resulting in a decrease in the lipid metabolic capacity of *Daphnia magna*. As a model zooplankton in an aquatic ecosystem, after exposure to algae-born silver, *Daphnia magna* can adjust its metabolic processes to adapt to the external environmental conditions. However, metabolic processes were regulated by enzyme activity and gene expression in the ordinary course of events. From the analyses, it can be concluded that the gene PLA2G (SPLA2), coding for the key enzyme secretory phospholipase A2, can not only be regarded as a biomarker but also be applied to the assessment of the aquatic food chain (e.g., *Scenedesmus obliquus* and *Daphnia magna*) exposed to metal-based nanoparticles.

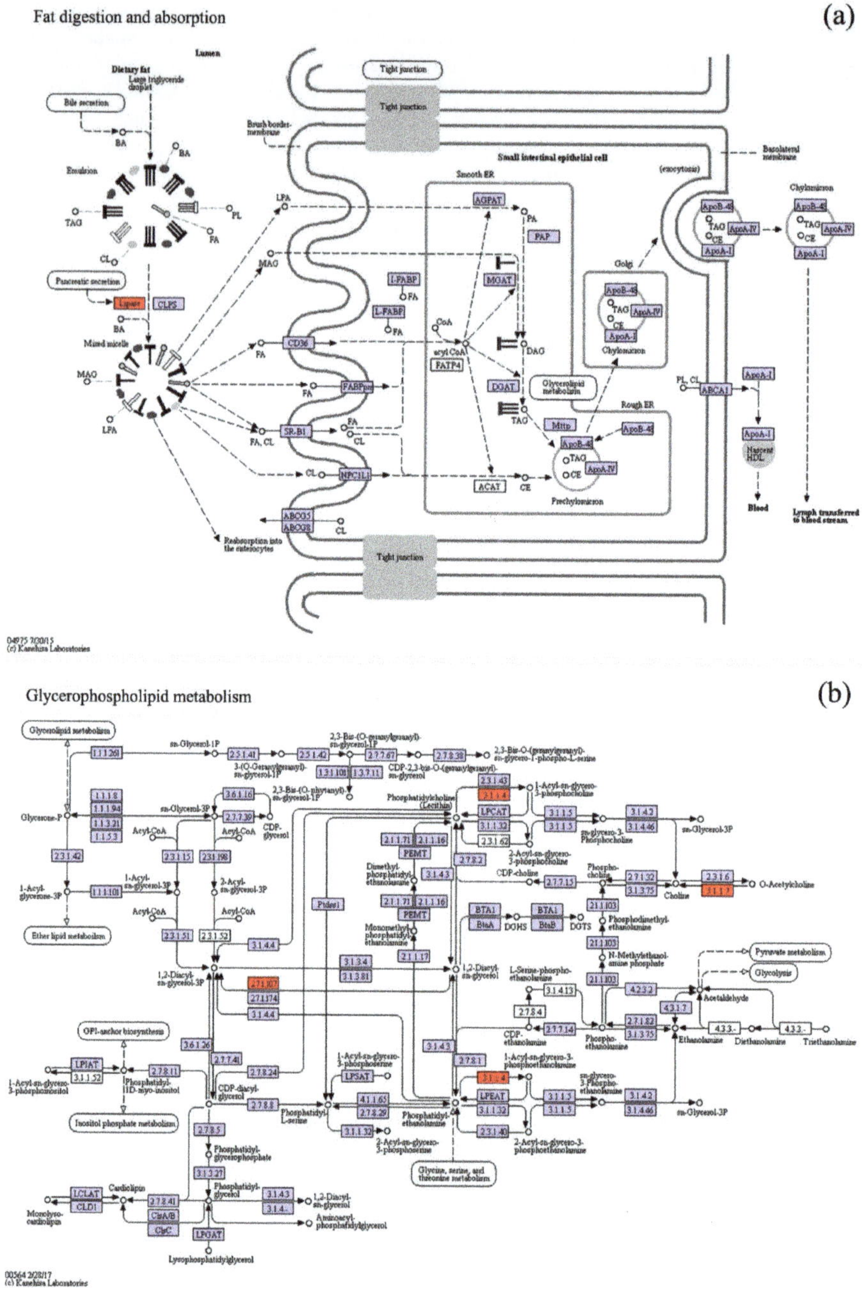

Figure 3. Effects of the diets of algae treated with AgNCs and silver ions on differentially expressed genes related to glycerophospholipid metabolism and fat digestion and absorption of *Daphnia magna*. Shades of red, light blue and white represent down-regulation of target unigenes, no differential expression of unigenes (background genes) and unigene products not involved in *Daphnia magna*, respectively. (**a**) Effects of the diets of algae treated with AgNCs (without L-cysteine) for 48 h on fat digestion and absorption of *Daphnia magna*. (**b**) Effects of the diets of algae treated with AgNCs

(without L-cysteine) for 48 h on glycerophospholipid metabolism of *Daphnia magna*. Based on the principles of classification of the Enzyme Commission, the identifier of the enzyme involved in the pathway was determined by the number in the boxes. Illustrations of the fat digestion and absorption and glycerophospholipid metabolism of *Daphnia magna* as a result of the diets of algae treated with AgNCs (containing L-cysteine) and silver ion are available in Table S5. Metabolic pathways were obtained through the KEGG annotation of differentially expressed unigenes.

Although the released Ag ions from AgNCs help to explain some of their toxicity, it is still not obvious whether AgNCs itself is the root of the problem. For example, Yuan et al. reported that AgNCs possessed superior antimicrobial properties against pathogenic bacteria (for example, *P. aeruginosa*), via the generation of intracellular reactive oxygen species (ROS) [33]. However, our previous work demonstrated that no significant ROS were induced in *Scenedesmus obliquus* treated with various concentrations of AgNCs and Ag ions for 96 h compared to the controls, suggesting that the AgNC- or Ag-ion-mediated toxicity to algae cells could not be attributable to ROS accumulation [34,35]. Another study reported that AgNCs had much higher antimicrobial activities towards both Gram-positive bacteria (*B. subtilis*, and *S. aureus*) and Gram-negative bacteria (*P. aeruginosa*, and *E. coli*) owing to abundantly released Ag ions on the surface [36]. With regard to the effect of their larger counterpart, AgNPs, Helen et al. reported that a 15k oligonucleotide microarray for *Daphnia magna* was used to differentiate between particle-specific and ionic silver toxicity [32]. AgNPs and $AgNO_3$ were found to have different gene expression profiles, which may indicate their different modes of toxicity. AgNPs and $AgNO_3$ can both affect key biological processes, including protein metabolism and signal transduction, as well as developmental processes (such as sensory development). A previous investigation of the toxicity of AgNPs in comparison with $AgNO_3$ in two aquatic organisms, green algae *Chlamydomonas reinhardtii* and the invertebrate *Daphnia magna*, revealed that *Daphnia magna* could equally accumulate AgNPs and $AgNO_3$ after feeding on Ag-containing algae, and the finding suggested that $AgNO_3$ was more toxic than nano-Ag [22]. Based on these previous studies and the significant changes in the expression levels of differentially expressed genes in *Daphnia magna* after exposure for 48 h to Ag-containing algae, we can confirm that the AgNCs and their released ionic-form silver have a synergistic effect on the invertebrate organism *Daphnia magna*. Whether the AgNCs entering the *Daphnia magna* cells will gradually degrade into silver ions and further cause its toxicity deserves further investigation. However, Chen and colleagues revealed that the toxicity of AgNPs on human monocytes (THP-1) is largely due to the chemical transformation of particulate silver from elemental silver $(Ag^0)_n$ to Ag^+ ions with the help of synchrotron radiation beam transmission X-ray microscopy (SR-TXM) and SR-X-ray absorption near edge structure (SR-XANES) spectroscopy, suggesting that this could be a potential mechanism [37].

4. Conclusions

In conclusion, we showed that *Daphnia magna* fed with *Scenedesmus obliquus* contained Ag-experienced transcriptome alterations as a result of exposure to AgNCs. AgNCs in their particulate form and the degraded Ag ions that they produce worked in concert to modify *Daphnia magna'* transcriptiome. Because the gene PLA2G (SPLA2), which codes for secretory phospholipase A2, was downregulated, the digestive process and glycerophospholipid metabolism were hindered. Additionally, it needs to be thoroughly researched in the future if the AgNCs that enter *Daphnia magna* cells gradually break down into Ag ions and further mediate AgNCs toxicity. The current study not only provides a different method for identifying AgNCs-mediated toxicity, but it also adds to our knowledge of early aquatic ecological risk mediated by other metal nanoparticles.

Supplementary Materials: The following supporting information can be downloaded at: https://www.mdpi.com/article/10.3390/w15183172/s1. Figure S1: The toxicity test of *Daphnia magna* through the diet of *Scenedesmus obliquus* treated with silver nanoparticles, and two photos represented the beginning (a) and end (b) of the toxicity test. Figure S2: The total RNA checked by agarose gel electrophoresis. Figure S3: Histogram of length distribution of contiges (a), transcripts (b) and unigenes (c). Figure S4: The Gene Ontology (GO) annotations of unigenes. Figure S5: The evolutionary genealogy of genes: non-supervised orthologous groups (eggNOG) annotations of unigenes. Figure S6: The Kyoto Encyclopedia of Genes and Genomes (KEGG) annotations of unigenes. Figure S7: The "volcano plot" pictures with $\log_2$(fold change) plotted versus $-\log_{10}$(p-value) of differentially expressed genes derived from *Daphnia magna*. Figure S8: GO enrichment analysis of differentially expressed genes of *Daphnia magna* from the diet of green algae. Table S1: The composition of SE culture medium. Table S2: The total RNA was checked using NanoDrop 2000 UV-Vis Spectrophotometer. Table S3: Illumina de novo assembly statistics of transcriptomic profiles of Daphnia magna from green algae diet. Table S4: Annotation of unigenes of *Daphnia magna* transcriptome. Table S5: List of differentially expressed genes for the digestive system and lipid metabolism of *Daphnia magna* after the diets of algae treated with silver for 48 h.

Author Contributions: Conceptualization, L.Z. and H.T.; methodology, H.T.; validation, L.Z. and H.T.; resources, H.T.; writing—original draft preparation, L.Z.; writing—review and editing, L.Z. and H.T.; visualization, L.Z.; funding acquisition, L.Z. All authors have read and agreed to the published version of the manuscript.

Funding: This work was supported by the Water Pollution Control and Treatment Projects of the Ministry of Science and Technology of China (2014ZX07206001), the National Natural Science fund of China (41272381) and the Campus for Research Excellence and Technological Enterprise (CREATE) programme between Singapore and Shanghai Jiao Tong University.

Data Availability Statement: Data are available from the corresponding author on request.

Acknowledgments: We appreciate Personal Biotechnology Co., Ltd. (Shanghai, China) for their supports in Transcriptome Sequencing.

Conflicts of Interest: The authors declare no conflict of interest.

References

1. Lu, Y.Z.; Chen, W. Sub-nanometre sized metal clusters: From synthetic challenges to the unique property discoveries. *Chem. Soc. Rev.* **2012**, *41*, 3594–3623. [CrossRef] [PubMed]
2. Diez, I.; Ras, R.H. Fluorescent silver nanoclusters. *Nanoscale* **2011**, *3*, 1963–1970. [CrossRef] [PubMed]
3. Shang, L.; Dong, S.J.; Nienhaus, G.U. Ultra-small fluorescent metal nanoclusters: Synthesis and biological applications. *Nano Today* **2011**, *6*, 401–418. [CrossRef]
4. Xu, H.X.; Suslick, K.S. Water-soluble fluorescent silver nanoclusters. *Adv. Mater.* **2010**, *22*, 1078–1082. [CrossRef] [PubMed]
5. Guidez, E.B.; Aikens, C.M. Theoretical analysis of the optical excitation spectra of silver and gold nanowires. *Nanoscale* **2012**, *4*, 4190–4198. [CrossRef] [PubMed]
6. Yuan, X.; Yeow, T.J.; Zhang, Q.; Lee, J.Y.; Xie, J.P. Highly luminescent Ag^+ nanoclusters for Hg^{2+} ion detection. *Nanoscale* **2012**, *4*, 1968–1971. [CrossRef]
7. Zheng, K.; Yuan, X.; Kuah, K.; Luo, Z.; Yao, Q.; Zhang, Q.; Xie, J.P. Boiling water synthesis of ultrastable thiolated silver nanoclusters with aggregation-induced emission. *Chem. Commun.* **2015**, *51*, 15165–15168. [CrossRef]
8. Murray, R.W. Nanoelectrochemistry: Metal Nanoparticles, Nanoelectrodes, and Nanopores. *Chem. Rev.* **2008**, *108*, 2688–2720. [CrossRef]
9. Yuan, X.; Tay, Y.; Dou, X.; Luo, Z.; Leong, D.T.; Xie, J.P. Glutathione-protected silver nanoclusters as cysteine-selective fluorometric and colorimetric probe. *Anal. Chem.* **2013**, *85*, 1913–1919. [CrossRef]
10. Guo, W.W.; Yuan, J.P.; Dong, Q.Z.; Wang, E. Highly sequence-dependent formation of fluorescent silver nanoclusters in hybridized DNA duplexes for single nucleotide mutation identification. *J. Am. Chem. Soc.* **2010**, *132*, 932–934. [CrossRef]
11. Yeh, H.C.; Sharma, J.; Han, J.J.; Martinez, J.S.; Werner, J.H. A DNA–silver nanocluster probe that fluoresces upon hybridization. *Nano Lett.* **2010**, *10*, 3106–3110. [CrossRef]
12. Zheng, C.R.; Li, S.; Ye, C.; Li, X.; Zhang, C.; Yu, X. Particulate respirators functionalized with silver nanoparticles showed excellent real-time antimicrobial effects against pathogens. *Environ. Sci. Technol.* **2016**, *50*, 7144–7151. [CrossRef]
13. Xiu, Z.M.; Zhang, Q.B.; Puppala, H.L.; Colvin, V.L.; Alvarez, P.J. Negligible particle-specific antibacterial activity of silver nanoparticles. *Nano Lett.* **2012**, *12*, 4271–4275. [CrossRef]
14. Leclerc, S.; Wilkinson, K.J. Bioaccumulation of nanosilver by *Chlamydomonas reinhardtii*-nanoparticle or the free ion? *Environ. Sci. Technol.* **2014**, *48*, 358–364. [CrossRef] [PubMed]

15. Kaveh, R.; Li, Y.S.; Ranjbar, S.; Tehrani, R.; Brueck, C.L.; Van Aken, B. Changes in *Arabidopsis thaliana* gene expression in response to silver nanoparticles and silver ions. *Environ. Sci. Technol.* **2013**, *47*, 10637–10644. [CrossRef] [PubMed]
16. Shao, Z.S.; Guagliardo, P.; Jiang, H.B.; Wang, W.X. Intra- and intercellular silver nanoparticle translocation and transformation in Oyster gill filaments: Coupling nanoscale secondary ion mass spectrometry and dual stable isotope tracing study. *Environ. Sci. Technol.* **2021**, *55*, 433–446. [CrossRef] [PubMed]
17. Gray, E.P.; Coleman, J.G.; Bednar, A.J.; Kennedy, A.J.; Ranville, J.F.; Higgins, C.P. 2013. Extraction and analysis of silver and gold nanoparticles from biological tissues using single particle inductively coupled plasma mass spectrometry. *Environ. Sci. Technol.* **2013**, *47*, 14315–14323. [CrossRef]
18. Liu, N.; Li, Y.; Liu, L.H.; Liu, X.L.; Yin, Y.G.; Qu, G.B.; Shi, J.B.; Song, M.Y.; He, B.; Hu, L.G.; et al. Administration of silver nasal spray leads to nanoparticle accumulation in rat brain tissues. *Environ. Sci. Technol.* **2022**, *56*, 403–413. [CrossRef]
19. Wang, M.Y.; Wang, W.X. Nanoscale whole-body expansion microscopy revealed the early skeletal developmental malformation induced by silver nanoparticles. *Environ. Sci. Technol. Lett.* **2023**, *10*, 471–477. [CrossRef]
20. Farre, M.; Gajda-Schrantz, K.; Kantiani, L.; Barcelo, D. Ecotoxicity and analysis of nanomaterials in the aquatic environment. *Anal. Bioanal. Chem.* **2009**, *393*, 81–95. [CrossRef]
21. Gupta, G.S.; Kumar, A.; Senapati, V.A.; Pandey, A.K.; Shanker, R.; Dhawan, A. Laboratory scale microbial food chain to study bioaccumulation, biomagnification, and ecotoxicity of cadmium telluride quantum dots. *Environ. Sci. Technol.* **2017**, *51*, 1695–1706. [CrossRef] [PubMed]
22. McTeer, J.; Dean, A.P.; White, K.N.; Pittman, J.K. Bioaccumulation of silver nanoparticles into Daphnia magna from a freshwater algal diet and the impact of phosphate availability. *Nanotoxicology* **2014**, *8*, 305–316. [CrossRef] [PubMed]
23. Quigg, A.; Chin, W.-C.; Chen, C.-S.; Zhang, S.; Jiang, Y.; Miao, A.J.; Schwehr, K.A.; Xu, C.; Santschi, P.H. Direct and Indirect Toxic Effects of Engineered Nanoparticles on Algae: Role of Natural Organic Matter. *ACS Sustain. Chem. Eng.* **2013**, *1*, 686–702. [CrossRef]
24. Chisholm, S.W. Stirring times in the Southern Ocean. *Nature* **2000**, *407*, 685–687. [CrossRef]
25. Yan, N.; Tang, B.Z.; Wang, W.X. In vivo bioimaging of silver nanoparticle dissolution in the gut environment of zooplankton. *ACS Nano* **2018**, *12*, 12212–12223. [CrossRef]
26. Levard, C.; Hotze, E.M.; Lowry, G.V.; Brown, G.E., Jr. Environmental transformations of silver nanoparticles: Impact on stability and toxicity. *Environ. Sci. Technol.* **2012**, *46*, 6900–6914. [CrossRef]
27. Lowry, G.V.; Gregory, K.B.; Apte, S.C.; Lead, J.R. Transformations of nanomaterials in the environment. *Environ. Sci. Technol.* **2012**, *46*, 6893–6899. [CrossRef]
28. Yu, S.J.; Yin, Y.G.; Liu, J.F. Silver nanoparticles in the environment. *Environ. Sci. Process. Impacts* **2013**, *15*, 78–92. [CrossRef]
29. Gil-Allue, C.; Schirmer, K.; Tlili, A.; Gessner, M.O.; Behra, R. Silver nanoparticle effects on stream periphyton during short-term exposures. *Environ. Sci. Technol.* **2015**, *49*, 1165–1172. [CrossRef]
30. Ankley, G.T.; Daston, G.P.; Degitz, S.J. Toxicogenomics in Regulatory Ecotoxicology. *Environ. Sci. Technol.* **2006**, *40*, 4055–4065. [CrossRef]
31. Chen, C.; Unrine, J.M.; Judy, J.D.; Lewis, R.W.; Guo, J.; McNear, D.H.; Tsyusko, O.V. Toxicogenomic responses of the model legume medicago truncatula to aged biosolids containing a mixture of nanomaterials (TiO_2, Ag, and ZnO) from a pilot wastewater treatment plant. *Environ. Sci. Technol.* **2015**, *49*, 8759–8768. [CrossRef] [PubMed]
32. Poynton, H.C.; Lazorchak, J.M.; Impellitteri, C.A.; Blalock, B.J.; Rogers, K.; Allen, H.J.; Loguinov, A.; Heckman, J.L.; Govindasmawy, S. Toxicogenomic responses of nanotoxicity in *Daphnia magna* exposed to silver nitrate and coated silver nanoparticles. *Environ. Sci. Technol.* **2012**, *46*, 6288–6296. [CrossRef] [PubMed]
33. Yuan, X.; Setyawati, M.I.; Tan, A.S.; Ong, C.N.; Leong, D.T.; Xie, J.P. Highly luminescent silver nanoclusters with tunable emissions: Cyclic reduction–decomposition synthesis and antimicrobial properties. *NPG Asia Mater.* **2013**, *5*, e39. [CrossRef]
34. Zhang, L.; He, Y.L.; Goswami, N.; Xie, J.P.; Zhang, B.; Tao, X.J. Uptake and effect of highly fluorescent silver nanoclusters on Scenedesmus obliquus. *Chemosphere* **2016**, *153*, 322–331. [CrossRef]
35. Zhang, L.; Goswami, N.; Xie, J.P.; Zhang, B.; He, Y.L. Unraveling the molecular mechanism of photosynthetic toxicity of highly fuorescent silver nanoclusters to *Scenedesmus obliquus*. *Sci. Rep.* **2017**, *7*, 16432. [CrossRef]
36. Yuan, X.; Setyawati, M.I.; Leong, D.T.; Xie, J.P. Ultrasmall Ag^+-rich nanoclusters as highly efficient nanoreservoirs for bacterial killing. *Nano Res.* **2013**, *7*, 301–307. [CrossRef]
37. Wang, L.; Zhang, T.; Li, P.; Huang, W.; Tang, J.; Wang, P.; Liu, J.; Yuan, Q.; Bai, R.; Li, B.; et al. Use of Synchrotron radiation-analytical techniques to reveal chemical origin of silver-nanoparticle cytotoxicity. *ACS Nano* **2015**, *9*, 6532–6547. [CrossRef]

water

Communication

Study on Plugging Material and Plugging Mechanism of Crude Oil Sand Water Filter Pipe

Wenhui Zhang [1], Qingfeng Liu [2], Hengyu Chen [1], Huibin Sheng [3], Jingen Yan [4], Yongtao Gu [4], Xianqiang Huang [5,*] and Bingchuan Yang [1,5,*]

1 College of Chemistry and Chemical Engineering, Qilu Normal University, Jinan 250013, China; zwh2963105101@163.com (W.Z.); qicechenyu@163.com (H.C.)
2 Shengli Oil Field Zhongsheng Petroleum Development Company, Dongying 257237, China; wolfmanliu007@163.com
3 Shandong Institute of Geophysical and Geochemical Exploration, Jinan 250013, China; helloshb@163.com
4 Gudong Oil Production Plant, Shengli Oilfield Company, China Petroleum Chemical Corporation, Dongying 257237, China; yanjingen@163.com (J.Y.); guyongtao@163.com (Y.G.)
5 School of Chemistry and Chemical Engineering, Liaocheng University, Liaocheng 252059, China
* Correspondence: hxq@lcu.edu.cn (X.H.); lewis19886@163.com (B.Y.)

Abstract: In order to develop the biological self-cleaning anti-clogging high-permeability sand filter tube, it is very important to analyze the plugging material and plugging mechanism of crude oil sand water filter. Under laboratory conditions, based on vacuum drying, condensation reflux, chromatographic separation, and other technologies, the plugging components were preliminarily analyzed. The plugging components were analyzed via XRD, infrared analysis, ^{1}H NMR, and ^{13}C NMR. Through analysis and testing, XRD results showed that the solid compositions were clay and sand grains. Meanwhile, the infrared analysis, ^{1}H NMR, and ^{13}C NMR demonstrated that the main components of the plug are asphaltene crude oil, and the proportion of aromatic components and saturated components is close.

Keywords: crude oil; sand water filter; plugging components; plugging mechanism

1. Introduction

Developing the loose sandstone hydrocarbon reservoirs, which account for a large proportion of China's oil and gas, is crucial [1]. However, frequent sanding from oil wells during development has hindered further development of the loose sandstone oil reservoirs, and the specific nature of the reservoir has been one of the main conflicts in the rapid development of the loose sandstone oil reservoirs [2,3]. As a result, it has become the primary problem in removing the sand out of the well.

The biggest difficulty encountered in the use of conventional sand filter pipes is clogging [4]. The blockage of the sand filter tubes can cause a reduction in the fluid output of oil well or even lead to the blockage of oil wells [5,6]. In addition, replacing the clogged sand filter pipes carries certain risks, as the pipes can easily fracture or break during replacement, leading to wellbore blockage and abandonment [7,8]. The biological self-cleaning anti-clogging high permeability sand filter tube, which has applied electrocatalytic technology, is a new direction for enhanced oil recovery [9–12]. However, clogging issues have prevented its widespread usage. Consequently, the investigation of crude oil clogging in sand filter pipes and the exploration of clogging mechanisms have emerged as crucial research areas for effective sand control in oil wells [13–15].

According to statistical data, the filling layer of polymer-aided flooding reservoirs is severely blocked, resulting in low liquid volume, rapid decrease in liquid volume, and a short effective period [16,17]. Conventional sand control methods have poor adaptability to post-polymer flooding reservoirs during high water cut periods. The main reasons for this include the exacerbation of particle transport caused by polymer carriage and drag, as well as the accumulation of aged polymers, which block the screen pipes and filling layer, leading to a rapid decrease in liquid volume and a shortened effective period [18]. Ma et al. presented a systematic approach to investigate the performance of sand control screens, including the processing of the sand control simulation unit, the analysis of formation parameters, the determining of pressure drop through the sand retaining medium, the calculation of oil-production index, and so on [19]. However, it does not solve the root cause of the clogging problem of clogged sand filter tubes. Therefore, in this paper, we analyzed the blockage mechanism and the blockages in the original oil filter sand pipes. It provides an effective avenue for addressing the problem of blockage of traditional sand control tools.

2. Oilfield Overview

Located in the Yellow River Delta region, Shengli Oilfield, the second largest oilfield in China is situated in the northern part of Shandong Province. Its main working area covers approximately 4.4×104 km^2. As a typical loose sandstone of a compound oil-bearing basin, Shengli Oilfield is characterized by complex geological conditions, a variety of oil-bearing layers, diverse types of reservoirs, and rich types of oil reservoirs, as well as significant variations in reservoir properties [20–22].

Shengli Oilfield has discovered 38 porous sandstone oilfields, with reserves of 3.716 billion tons and an annual output of 23.244 million tons, accounting for more than 80% of the oilfield's reserves and annual output. As a fundamental technology for developing porous sandstone reservoirs, sand control technology is the dominant technology for ensuring the normal production of such reservoirs. With an average annual construction of 2365 wells, accounting for more than 20% of the oilfield's measures, sand control plays a critical role in ensuring sustained and stable production of the oilfield [23,24].

3. Experimental Part

3.1. Experimental Reagent

Dichloromethane (analytical pure), petroleum ether (analytical pure), toluene (analytical pure), and alumina (column chromatography grade) were purchased from Shanghai in China. The water is primarily distilled water.

3.2. Experimental Instrument

The Fourier transform infrared spectrometry (FT-IR) analyses were measured on a Nicolet 5700 spectrophotometer in the range of 400–4000 cm^{-1}. ^{1}H NMR and ^{13}C NMR spectra were recorded on a 400 MHz (Bruker Company in German, Bremen Germany) instrument, and $CDCl_3$ (7.26 ppm for ^{1}H NMR, 77.16 ppm for ^{13}C NMR) was used as a reference. The X-ray diffraction (PXRD) patterns were recorded on a Rigaku Smartlab3 X-ray Powder Diffractometer (Rigaku Company in Japan, Tokyo, Japan) equipped with a Cu-sealed tube (λ = 1.54178 Å) in the range of 5° to 50° at room temperature. Microscope visual rheometer (MCR302, Anton Paar, Graz, Steiermark, Austria).

3.3. Crude Oil Composition Analysis

Crude oil samples from wells in different oil blocks were selected for compositional analysis: well 641X15 in Block 641, well 822P5 in Block 821, well 827P1 in Block 827, well 9P13 in Block 9, and well 10N14 in Block 4. Crude oil samples are shown in Figure 1.

Figure 1. Crude oil samples from the crude oil zone.

4. Analysis of Experiment

Approximately 3 g samples were taken, respectively (as shown in Figure 2), and dried under vacuum at 110 °C and −1000 KPa for 1 h after cooling and weighing (accurate to 0.001 g). For every 1 g of the sample, 30 mL of crude ether was added, boiled, and refluxed in a reflux condenser. The sample was then filtered after standing in a dark condenser at room temperature for 1.5–2.5 h. The insoluble material was extracted with the filtrate, and the liquid at the bottom of the extractor was removed, placed on filter paper, and allowed to evaporate. If there was no residue on the filter paper, a solution of the soluble matter was obtained.

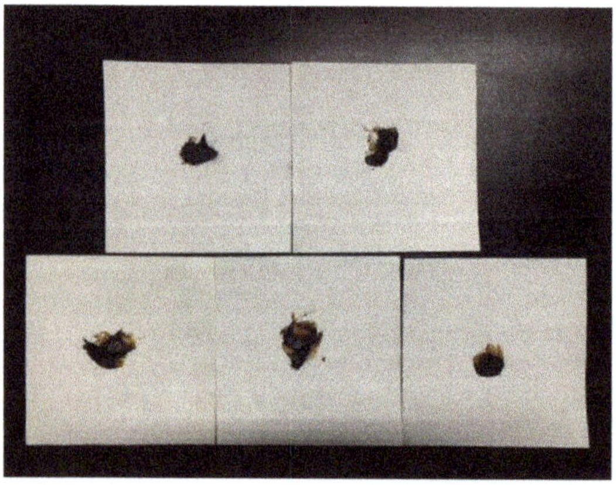

Figure 2. Crude oil samples to be tested in different oil areas.

The extraction of the filter paper continued according to the proportion of 30 mL of toluene added to each 1 g of the sample to obtain the asphalt solution. After vacuum drying the soluble component solution with 30 mL of crude ether, it can be separated into three components using a chromatographic column containing alumina in the following order: saturated components, aromatic components, and resin components. After evaporating most of the solvent from the component solution, it was dried under vacuum at 110 °C and −1000 KPa for 1 h, cooled to room temperature, and weighed, and the content of all components was calculated.

4.1. Experimental Results

The experiments showed that the crude oil from the Gudong Oil Production Area of the Shengli Oilfield has a high content of gums and asphaltenes, and the aromatic and saturated fractions are close to each other.

Due to the high gum and asphaltene content, crude oil is easily agglomerated to form a high-viscosity crude oil block while adhering to the surface of pipes, containers, and rock formations. At the same time, due to the strong binding ability, crude oil block partially wrapped in stratum sand and clay, forming inorganic and organic compounds difficult to decompose soft scale block.

Through an NMR analysis of crude oil and blockages, it can be determined that the main component of the crude oil blockage is the asphalt and colloidal components of the crude oil, which have strong adhesion capabilities and cause pore blockage in filter sand pipes, resulting in a gradual reduction in the production liquid over time. We confirmed the proportion of the component of crude oil through five repetitions of the experiment and listed the results of the averages in Table 1.

Table 1. Analysis of crude oil composition of different wells in Gudong Oil Production Plant.

Well Number	Aromatic Components%	Saturated Component%	Gum%	Asphaltene%	Gum + Asphaltene%
641-X15	32.1	31.5	28.9	7.5	36.4
822-P5	28.9	29.4	33.5	8.2	41.7
827-P1	31.4	28.9	32.9	6.8	39.7
9-P13	26.5	27.3	38.3	7.9	46.2
10-N14	33.6	32.7	25.9	7.8	33.7

4.2. Composition Analysis of Plugging Materials

The samples of pipeline and screen pipe blockages taken from the injection–production wells for sand control in Gudong Oilfield are shown in Figures 3 and 4. In order to analyze the composition of clogging deposits, 3 g of the clogging deposits were weighed and vacuum dried at 110 °C and −1000 KPa for 1 h, followed by cooling and weighing (accurate to 0.001 g).

Figure 3. Oil production pipeline blockage.

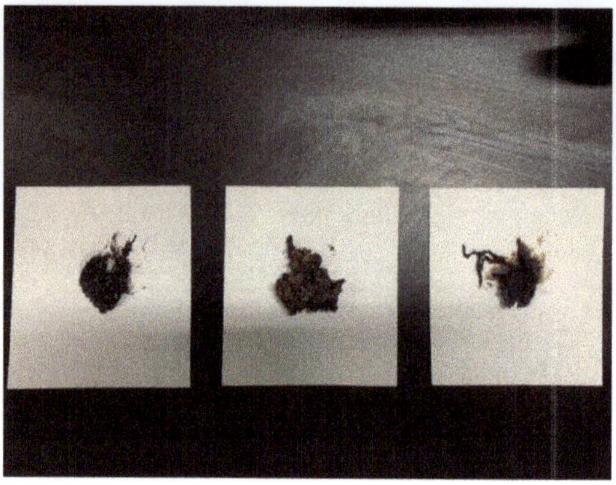

Figure 4. Sand filter pipe blockage.

First, 30 mL of petroleum ether was added to each 1 g sample, which was then boiled and refluxed in a reflux condenser, after which it was allowed to stand for 1.5–2.5 h at room temperature in a dark box. After filtration, the insoluble material was extracted until the liquid at the bottom of the extractor was removed. The liquid is applied to the filter paper and allowed to evaporate. If no residue remains on the filter paper, the result is a soluble solution. The saturated hydrocarbons, aromatic hydrocarbons, and gums from the stopper were separated. Continuing with the same 1 g sample to 30 mL toluene ratio, the filter paper was further extracted, obtaining an asphaltene solution.

The asphaltene in clogging deposits was separated, and clay and sand grains were separated by adding distilled water. The XRD results (Figure 5) showed that the main component of clogging deposits was crude oil (excluding asphaltene), accounting for 61.0%, followed by clay and formation sand, accounting for 20.8%, and asphaltene, accounting for 18.2%. XRD analysis was used to determine the composition of clay and sand grains, which contained 36% quartz (α-SiO_2), 21% plagioclase ($CaAl_2Si_2O_8$), 16% forsterite (Mg_2SiO_4), 11% aluminosilicate (Al_2SiO_5), 9% pyroxene ($NaAlSi_3O_8$), and 7% clay. The main components are all SiO_2 and silicates.

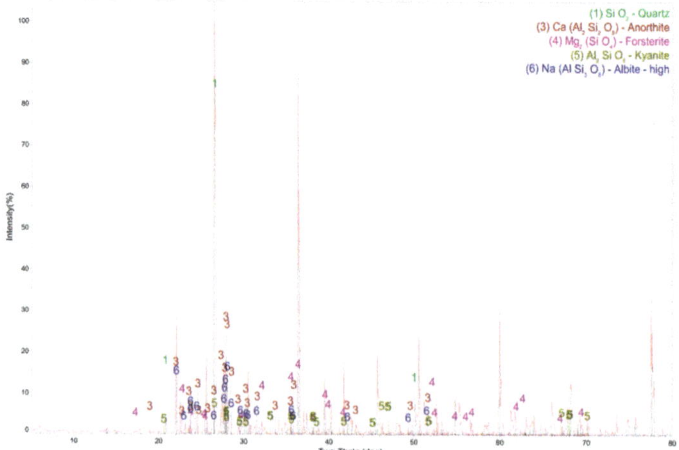

Figure 5. Blockage XRD powder diffraction pattern.

In addition, the blockage sample was subjected to infrared and nuclear magnetic resonance ^{1}H NMR and ^{13}C NMR tests and the results were shown in Figures 6–8. An analysis of the test results confirmed that the main component of the blockage was crude oil containing asphaltene, with a similar ratio of aromatic components to saturated components.

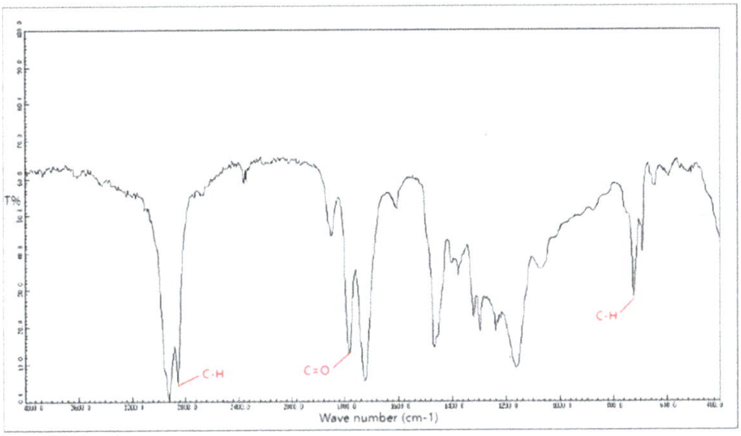

Figure 6. Infrared test of oil pipeline blockage.

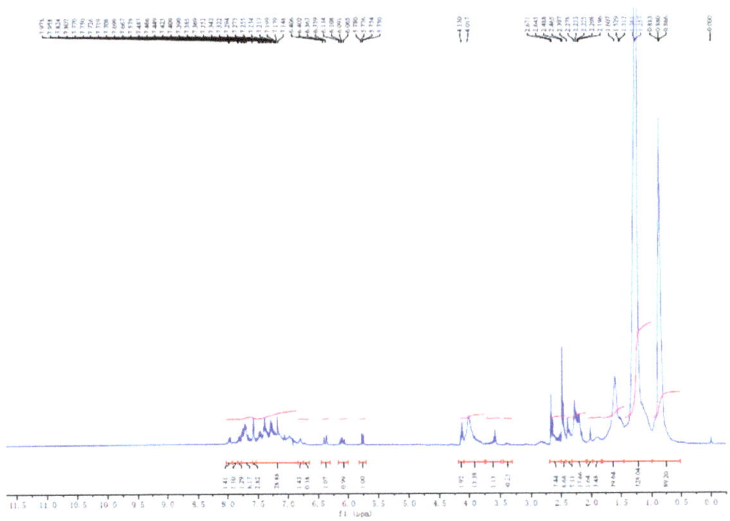

Figure 7. Oil production pipeline blockage ^{1}H NMR test results.

The results of the testing of the clogging deposits in the oil pipeline show that it is a mixture of saturated hydrocarbons, unsaturated alkenes, and aromatic hydrocarbons, as revealed in the NMR spectrum. The infrared spectrum shows a vibration peak of double bonds at 1800, indicating the presence of unsaturated hydrocarbons, and the characteristic C-H of aromatic ring absorption peak at 750 cm^{-1}. The hydrogen spectrum of the mixture shows saturated hydrogen atoms at 1–2.5 chemical shifts, olefinic hydrogen atoms at 3.5–4.5, and aromatic hydrogen atoms at 6.0–8.0. The carbon spectrum of the mixture confirms this, with many saturated hydrocarbon peaks below 100 chemical shifts and many aromatic and unsaturated hydrocarbon peaks at 120–145.

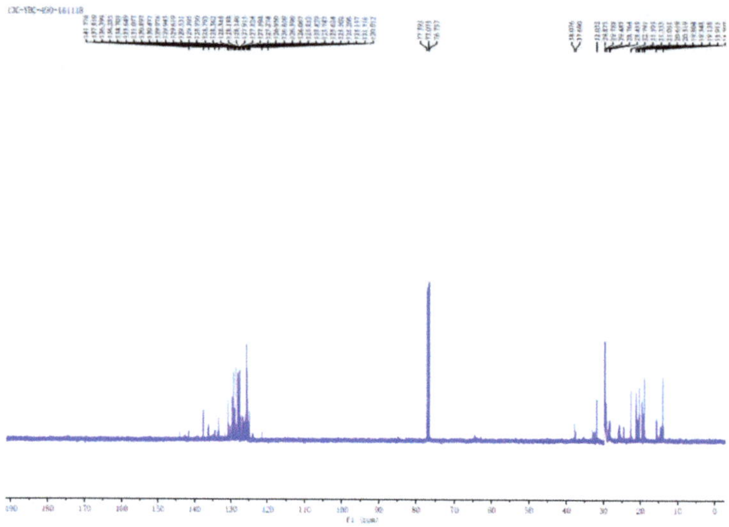

Figure 8. ^{13}C NMR test results of oil production pipeline blockage.

5. Analysis of Clogging Mechanism

5.1. With the Decrease of Temperature, the Viscosity of Crude Oil

After the crude oil well enters the production period, with the passage of time, the formation temperature decreases, and the viscosity of the crude oil increases. The viscosity–temperature relationship equation is as follows:

$$\ln\ln(Uo + 1) = A - B.\ln(t)$$

As the viscosity of crude oil increases, it will increase the near-well flow resistance, making it easier for the resin and asphaltene components in the crude oil to coalesce and adhere to the surface of sand control tools. The results in Figure 9 were investigated using a microscope visual rheometer (MCR302, Anton Paar, Graz, Steiermark, Austria).

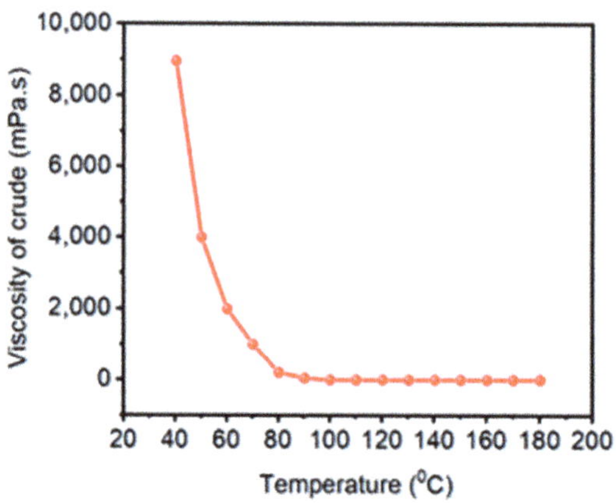

Figure 9. Crude oil viscosity curve with temperature.

5.2. Good Wettability of Crude Oil to Stainless Steel

When different phases of substances come into contact, assuming that there is no contact surface, there is no free surface energy. On the contrary, due to the interaction between molecules of each phase, free surface energy can be formed between the contact surfaces of the two phases. The larger the interface between the two phases, the greater the free surface energy. The distribution of fluids in the rock surface and the size and direction of pore capillary force in the reservoir will be directly influenced by the interfacial tension in the reservoir. Thus, the fluid flow in the reservoir will be further affected.

According to the second law of thermodynamics, substances always tend to decrease any of their free energy. Therefore, it is usually achieved by reducing the contact area, such as mercury shrinking into a spherical shape on a desktop, adsorbing molecules of adjacent contact substances, and using wetting to reduce the free energy of the system. In the pores of the rock, the microscopic distribution of oil, gas, and water is influenced by wetting. The wetting phenomenon in the reservoir usually occurs in the small capillaries of the rock. Liquid forms a curved shape in the capillary, thus generating a capillary force, which hinders fluid flow. The capillary force always acts on the non-wetting phase side.

During the water injection process in the reservoir, if the reservoir rock is hydrophilic, the oil can be lifted by the water; if the rock is hydrophobic, a layer of oil will be adsorbed around the pores, and the water can only drive the oil in the middle. Therefore, when the surface of the reservoir rock is hydrophilic, the capillary force is the driving force for water flooding; when the surface of the rock is hydrophobic, the capillary force is the resistance to water flooding. In addition, the migration ability (sensitivity) of particles in the reservoir is also affected by wettability. In the reservoir, particles wetted by the fluid are easily migrated with the fluid, while particles that are not soaked are not easily migrated.

Under certain conditions, the hydrophilicity or hydrophobicity of the sand surface of a filter sand tube can be converted, and this process is called the reversal of wettability. For example, quartz sand is mainly composed of silicates and generally shows hydrophilicity. However, if its surface adsorbs active substances, it can be changed to hydrophobicity, and water does not easily drive out oil. Advanced fatty acids are usually surface-active substances in crude oil. Thus, wettability reversal occurs in the sand of the filter sand tube under the action of surface-active substances. The local action and reversal can cause non-uniform wettability. Therefore, different wettability is exhibited in pores of different sizes (Figure 10).

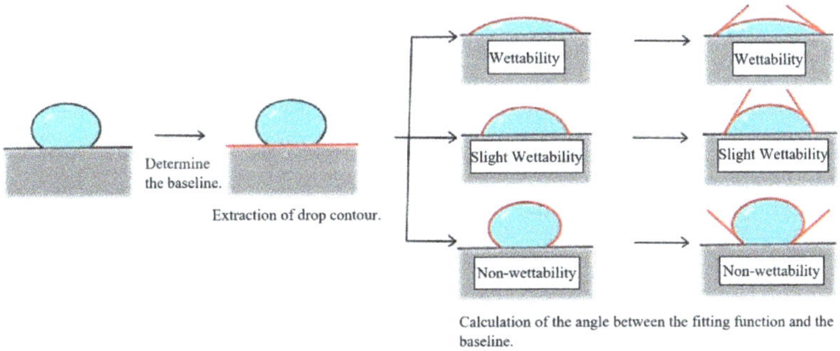

Figure 10. Contact wetting of substances with different properties.

Due to the strong affinity for crude oil, common metal filter sand tubes have a small wetting angle of crude oil on the metal surface, which makes it easy for crude oil to adsorb and aggregate on the surface. Crude oil contains a large amount of colloidal asphaltene, which makes it easier to agglomerate with clay and sandy layers and change the wettability from hydrophilicity to hydrophobicity, leading to a decrease in the relative permeability of the oil phase in the reservoir rock. In addition, organic deposits will also accumulate on the surface of asphaltene pores, causing the blockage of the reservoir pore throats (Figure 11).

Figure 11. Crude oil agglomerates and adsorbs stainless steel wire winding screen tubes.

5.3. Low Degree Polymerization of Aromatic Hydrocarbon and Unsaturated Hydrocarbon Occurred in Crude Oil

Comparing the infrared and nuclear magnetic resonance spectra of crude oil and blockages, it can be seen that crude oil contains a large number of saturated hydrocarbons and a small amount of unsaturated hydrocarbons and hydrocarbons. Four-zone blockage ^{1}H NMR and ^{13}C NMR test results and nine-zone blockage ^{1}H NMR and ^{13}C NMR test results were shown in Figure 12. 9P13, 10N14, 641X5, 822P5, 827P1 well crude oil ^{1}H NMR and ^{13}C NMR test results were shown in Figure 13. However, there are few unsaturated hydrocarbons in the blockages, and the proportion of aromatic hydrocarbons increases, shown in Figure 12, yellow box, ^{1}H NMR $\delta = 7.0$–8.0. This proves that under high temperatures, low-degree polymerization of unsaturated hydrocarbons occurs on the surface of the filter sand tube(shown in Figure 13, green box, ^{1}H NMR $\delta = 4.5$–5.5), resulting in low polymerization degree long-chain alkanes and some aromatic hydrocarbons, which, in turn, cause blockage deposits. All the details are shown in Supplementary Materials.

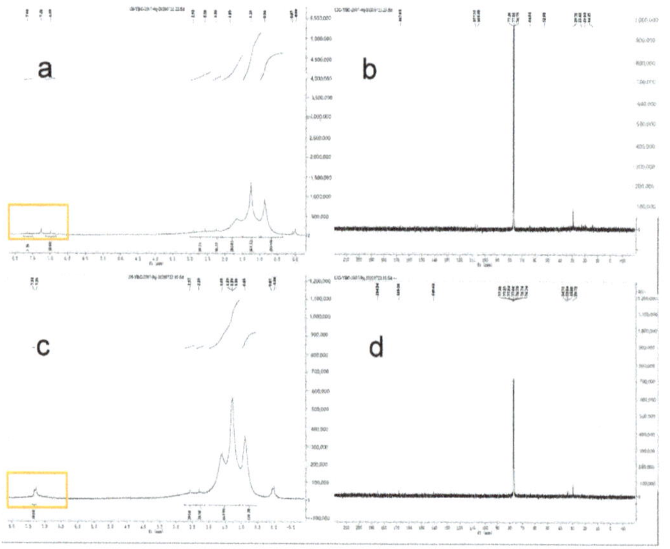

Figure 12. (**a**,**b**) Four-zone blockage ^{1}H NMR and ^{13}C NMR test results; (**c**,**d**) nine-zone blockage ^{1}H NMR and ^{13}C NMR test results.

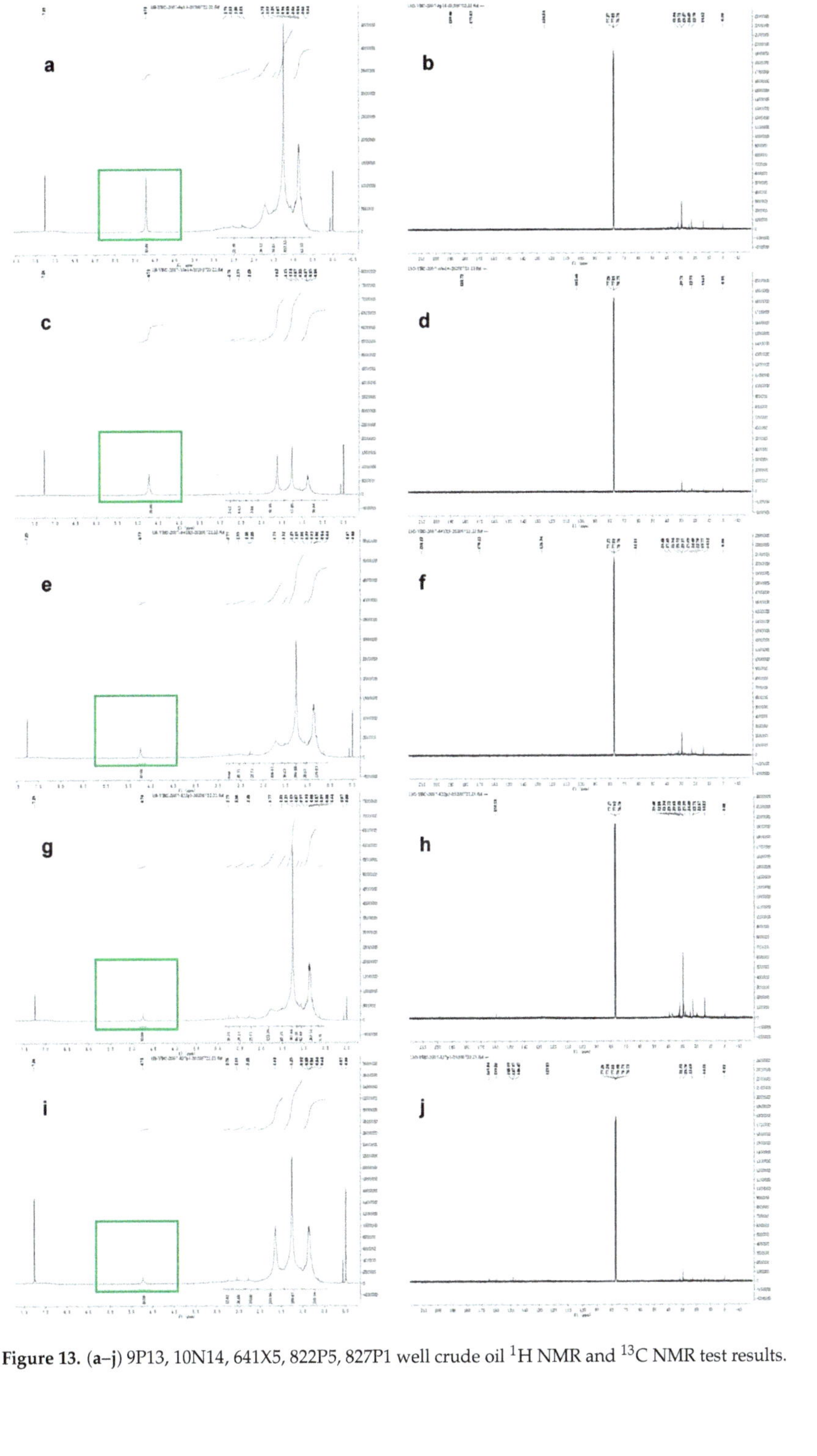

Figure 13. (**a**–**j**) 9P13, 10N14, 641X5, 822P5, 827P1 well crude oil ^{1}H NMR and ^{13}C NMR test results.

6. Conclusions

This article analyzes the main components of the crude oil in the Gudong oil production area of the Shengli oilfield and the main components of the blockage via on-site sampling. By comparing nuclear magnetic resonance and infrared analysis, we found that the blockage is mainly composed of asphaltenes and resins in crude oil. The asphaltenes and resins stick to the formation sand to form an indissoluble soft scale, which leads to the blocking of the filter sand pipe pores.

The laboratory analysis of the composition of the blockage sample reveals that the main component is crude oil, followed by substances such as clay and formation sand. XRD analysis of the clay and formation sand determines that their main components are silicon dioxide and silicate. Through the nuclear magnetic resonance and infrared analysis of the blockage sample, we determined that its main component is petroleum-containing asphaltene, and a mixture of saturated hydrocarbons, unsaturated olefins, and aromatic hydrocarbons is present.

Comparing the nuclear magnetic resonance of the crude oil and the blockage for the first time, we discovered that the unsaturated components in the crude oil polymerize on the pipeline surface, producing low polymers, which cause pipeline blockages. Through pipeline wettability analysis, we inferred that the pipeline blockage is severe, and the infiltration of crude oil on the pipeline leads to pipeline blockages.

Supplementary Materials: The following supporting information can be downloaded at: https://www.mdpi.com/article/10.3390/w15213714/s1. Figure S1: Blockage PXRD powder diffraction pattern; Figure S2: FT-IR spectra of oil pipeline blockage; Figure S3: Oil production pipeline blockage 1H NMR test results; Figure S4: Oil production pipeline blockage 13C NMR test results; Figure S5: 4-zone blockage 1H NMR test results; Figure S6: 4-zone blockage 13C NMR test results; Figure S7: 9 zone blockage 1H NMR test results; Figure S8: 9 zone blockage 13C NMR test results; Figure S9: 9P13 well crude oil 1H NMR test results; Figure S10: 9P13 well crude oil 13C NMR test results; Figure S11: 10N14 well crude oil 1H NMR test results; Figure S12: 10N14 well crude oil 13C NMR test results; Figure S13: 641X5 well crude oil 1H NMR test results; Figure S14: 641X5 well crude oil 13C NMR test results; Figure S15: 822P5 well crude oil 1H NMR test results; Figure S16: 822P5 well crude oil 13C NMR test results; Figure S17: 827P1 well crude oil 1H NMR test results; Figure S18: 827P1 well crude oil 13C NMR test results.

Author Contributions: Investigation, Q.L. and H.C.; Methodology, H.S. and J.Y.; Project administration, Y.G.; Supervision, B.Y. and X.H.; Writing—original draft, W.Z. All authors have read and agreed to the published version of the manuscript.

Funding: We are grateful to the National Innovation and Entrepreneurship Training Program for College Students (CXCY2023105, S202210447044, 202210447015) for the financial support of this research.

Data Availability Statement: The data used to support the findings of this study are available from the corresponding author upon request.

Conflicts of Interest: The authors declare no conflict of interest.

References

1. Liu, J.; Ji, M.; Qin, J. Catalytic Performance of Fe-Rich Sludge in Pyrolysis of Waste Oil Scum as Volatiles and Magnetic Char. *Water* **2023**, *15*, 2637.
2. Ikporo, B.; Sylvester, O. Effect of sand invasion on oil well production: A case study of Garon field in the Niger Delta. *Int. J. Eng. Sci.* **2015**, *4*, 64–72.
3. Naseer, M.T.; Khalid, R.H.; Naseem, S. Static Reservoir Simulations and Seismic Attributes Application to Image the Miocene Deep-Water Reservoirs in Southeast Asia. *Water* **2023**, *15*, 2543.
4. Sanni, S.E.; Adefila, S.S.; Anozie, A.N. Prediction of sand kinematic pressure and fluid-particle interaction coefficient as means of preventing sand-induced corrosion in crude oil pipelines. *Ain Shams Eng. J.* **2019**, *10*, 55–62.
5. Ding, B.; Dong, M. Optimization of plugging high mobility zones in oil sands by injection of oil-in-water emulsion: Experimental and modeling study. *Fuel* **2019**, *257*, 116024.
6. Hua, Z.; Lin, M.; Dong, Z. Study of deep profile control and oil displacement technologies with nanoscale polymer microspheres. *J. Colloid Interface Sci.* **2014**, *424*, 67–74.

7. Zhang, Z.; Cao, G.; Zuo, J. Rheological properties and plugging behavior of active crude oil. *Pet. Sci. Technol.* **2020**, *38*, 131–145.
8. Gao, X.; Huang, Q.; Zhang, X. Estimating wax plug transportation force in crude oil pipeline pigging. *Energy Fuels* **2020**, *34*, 3110–3120.
9. Yang, B.; Ma, S.; Cui, R. Simultaneous removal of NOx and SO_2 with H_2O_2 catalyzed by alkali/magnetism-modified fly ash: High efficiency, low cost and catalytic mechanism. *Chem. Eng. J.* **2019**, *359*, 233–243.
10. Huang, X.; Gu, X.; Zhang, H. Decavanadate-based clusters as bifunctional catalysts for efficient treatment of carbon dioxide and simulant sulfur mustard. *J. CO_2 Util.* **2021**, *45*, 101419.
11. Yang, B.; Ma, S.; Cui, R. Novel low-cost simultaneous removal of NO and SO_2 with ·OH from decomposition of H_2O_2 catalyzed by alkali-magnetic modified fly ash. *Ind. Eng. Chem. Res.* **2019**, *58*, 5339–5347.
12. Huang, X.; Gu, X.; Qi, Y. Decavanadate-based Transition Metal Hybrids as Bifunctional Catalysts for Sulfide Oxidation and C—C Bond Construction. *Chin. J. Chem.* **2021**, *39*, 2495–2503.
13. Sanni, S.E.; Adefila, S.S.; Anozie, A.N. Mechanisms for Controlling Sand-Induced Corrosion in Horizontal Pipe Flow of Sand, Crude Oil and Water. *Open Pet. Eng. J.* **2017**, *10*, 220–238.
14. Dong, W.; Zhang, D.; Wang, K. Investigation on degradation mechanism of polymer blockages in unconsolidated sandstone reservoirs. *e-Polymers* **2020**, *20*, 55–60.
15. Wang, F.; Yang, H.; Jiang, H. Formation mechanism and location distribution of blockage during polymer flooding. *J. Pet. Sci. Eng.* **2020**, *194*, 107503.
16. Nikoo, A.H.; Ghaedi, M.; Malayeri, M.R. Analysis of wellbore clogging by asphaltene deposition using interaction energies. *Fuel* **2023**, *352*, 129111.
17. Dong, C.; Jia, B.; Liu, C. Blocking mechanism and blocking laws experiments of sand retention media in mechanical screens. *J. China Univ. Pet. Ed. Nat. Sci.* **2011**, *35*, 82–88.
18. Ma, C.; Feng, Y.; Deng, J. Experimental investigation of the plugging mechanisms of non-consolidated prepacked gravel screens. *Int. J. Hydrogen Energy* **2021**, *46*, 34638–34651.
19. Ma, C.; Deng, J.; Dong, X. A new laboratory protocol to study the plugging and sand control performance of sand control screens. *J. Pet. Sci. Eng.* **2020**, *184*, 106548.
20. Liu, D.; Liu, H.; Li, L. Corrosion of water injection system in Shengli Oil Field. *Anti-Corros. Methods Mater.* **2013**, *60*, 185–193.
21. Li, C.-F.; Li, Y.; Li, X.-M.; Cao, Y.-B.; Song, Y.-T. The application of microbial enhanced oil recovery technology in Shengli Oilfield. *Pet. Sci. Technol.* **2015**, *33*, 556–560. [CrossRef]
22. Yu, Y.; Zhang, W.; Chen, G. Preparation of petroleum-degrading bacterial agent and its application in remediation of contaminated soil in Shengli Oil Field, China. *Environ. Sci. Pollut. Res.* **2014**, *21*, 7929–7937. [CrossRef] [PubMed]
23. Liu, J.; Song, X.; Sun, R. A Petroleum pollution and the microbial community structure in the soil of Shengli Oilfield. *Yingyong Shengtai Xuebao* **2014**, *25*, 850–856. [PubMed]
24. Wu, X.; Li, S.; Yan, F. The Application of Cluster Analysis Method in Reservoir Evaluation of Qingdong 5 Block in Shengli Oilfield. *Adv. Geosci.* **2017**, *7*, 135–141. [CrossRef]

MDPI AG
Grosspeteranlage 5
4052 Basel
Switzerland
Tel.: +41 61 683 77 34

Water Editorial Office
E-mail: water@mdpi.com
www.mdpi.com/journal/water

www.ingramcontent.com/pod-product-compliance
Lightning Source LLC
Chambersburg PA
CBHW062034220926
51850CB00010B/2544